Past and Future Freshwater Use in the United States: A technical document supporting the 2000 USDA Forest Service RPA Assessment

Thomas C. Brown

Contents

Introduction

Water withdrawals to cities, farms, and other offstream uses in the United States have increased over ten-fold during the twentieth century in response to tremendous population and economic growth. Further rapid growth in population and income is almost certain to occur, placing additional demands on water supplies. As withdrawals to offstream users increase, more water is consumed, leaving less water in streams. Streamflows have dropped at the same time as additional instream uses have been found by scientists studying the needs of aquatic plants and wildlife and the hydro-geologic requirements of river channels themselves, and as rising incomes and urbanization have intensified calls for maintaining water-based recreation opportunities and protecting water quality (Gillilan and Brown 1997). These changes amplify the importance of examining the future adequacy of the nation's water supply. As Congress recognized when it passed the Forest and Rangeland Renewable Resources Planning Act of 1974 requiring the Forest Service to periodically assess anticipated resource supply and demand conditions, with sufficient forethought necessary adjustments may be anticipated and unnecessary costs may be avoided.

The adequacy of a water supply depends on water availability compared with water demand. This report focuses on water demand, and estimates future water use assuming that the water will be available. Comparison of water-use estimates presented in this report with estimates of future water availability is left to a later report.

In economic terms, demand is a price-quantity relation. Unfortunately, such relations are difficult to specify for some water uses and for large geographic regions containing numerous market areas. Thus, an economic model was not adopted for this study. Instead, demand, as used in this report, refers to quantity requested. This quantity-based approach leaves the effect of price unspecified but not avoided. Because water and the resources needed to manage it are scarce, price has played an important role in determining the past quantities of water requested and will continue to do so. In what follows, the implicit role of price must be remembered.

Demand for water differs by region. Arid areas have higher demands per user than do humid areas, all else equal. Within a region of homogeneous weather, demands differ geographically depending on the availability of arable land, reliance on thermoelectric power, and other factors. The many potential differences among geographic areas suggest that demand for water should be studied at the smallest geographical scale possible. However, existing small-scale studies, often performed using different variables or methods, do not lend themselves to broad-scale conclusions about regional or national trends.

Large-scale projections of water use in the U.S. were attempted in 1961 by the Senate Select Committee on National Water Resources, in 1971 by Wollman and Bonem for Resources for the Future, in 1968 and 1978 by the Water Resources Council, in 1973 by the National Water Commission, and in 1989 by Guldin of the USDA Forest Service. Comparisons of these forecasts have consistently found large differences among them in projected water use, and large discrepancies between projected and actual water use (Viessman and DeMoncada 1980, Osborn and others 1986, Guldin 1989). These differences highlight the dangers of extrapolation and forecasting without a detailed understanding of the determinants of water use (Shabman 1990).

However, knowing the determinants of water use and how they interact is only half the job of forecasting resource use. Accurate forecasts also require accurate estimates of future determinant levels. Without the ability to accurately forecast future levels of all independent variables, increasing model complexity by adding variables to more accurately characterize past use may complicate the forecasting effort, not enhance it.

Guessing about future water use is like most other attempts to divine the future: the only thing we are quite sure of is that the future will not turn out as we expect. Accurate forecasts of future water use are impossible because we know too little about future technological and economic conditions. Thus, we must lower our expectations. What is possible is to project water demand assuming a continuation of recent past trends in factors that affect water use. Estimates of future possibilities based on projecting past trends offer a starting point for considering possible adjustments in water prices, management facilities, and institutions. This study emphasizes projections based on major water-use determinants (population, income, electric energy production, irrigated acreage) considering information on 1960 through 1995 trends in water-use efficiency. Recognizing the difficulty of forecasting, the overall approach I take is to minimize complexity so that underlying assumptions are relatively few, and their impact on the results is obvious.

This report projects water demand to the year 2040. The time horizon was selected based on the Forest and Rangeland Renewable Resources Planning Act, which mandates that the Forest Service periodically prepare a management plan for a period of roughly 45 years into the future. Of course, the likelihood that a projection is accurate decreases as the time horizon of that projection increases.

The objective of this paper is to characterize past and future water use in the U.S. A national perspective is first adopted to present a basic understanding of water-use trends. Then water use is described for large regions of the U.S. to capture the major regional differences.

Large Scale Water Use Data: Sources and Definitions

Except for some early estimates of water use from the Census Bureau, this report relies on water-use data from the U.S. Geological Survey (USGS). The USGS has estimated the nation's water use at five-year intervals since 1950. These periodic reports, issued as the following USGS circulars, represent the only consistent effort to document water use for the nation: MacKichan (1951, 1957), MacKichan and Kammerer (1961), Murray (1968), Murray and Reeves (1972, 1977), and Solley and others (1983, 1988, 1993, 1998). The circulars cover instream use at hydroelectric plants, withdrawals for delivery to offstream locations such as farms and cities, and consumptive use, which is the portion of a withdrawal that evaporates, transpires, or is incorporated into an end product, becoming unavailable for use by others within the basin. The portion of the withdrawal that is not consumptively used either returns to the stream (return flow) or contributes to groundwater storage. Of the offstream measures, the 1950 and 1955 circulars estimated only water withdrawal, but since 1960 both withdrawal and consumptive use have been estimated. Because of an interest in both withdrawal and consumptive use, this paper uses data found in the USGS circulars published since 1960. Including the most recent data, for 1995, the circulars used provide estimates for eight separate years covering a 35-year span.

The USGS estimates water use from three principal sources: groundwater, fresh surface water, and saline surface water. This report focuses on freshwater use. Unless stated otherwise, withdrawals and consumptive use are presented for the combination of ground and surface water.

Since 1955, the USGS water-use circulars have estimated use for the nation's major watersheds or large areas of contiguous coastal watersheds, called water resource regions (WRRs) (Water Resources Council 1978). Before 1970, the circulars aggregated data from the Texas-Gulf region and the Rio Grande region (WRRs 12 and 13) into what was called the Western Gulf region, and combined the Upper and Lower Colorado regions (WRRs 14 and 15) into a combined Colorado region. To have consistent regions for all years, the 1960 and 1965 data for these two larger regions were allocated to their respective WRRs based on the proportions of population that resided in the separate WRRs at that time.

Water use is summarized in this report first for the United States as a whole and then for the 20 WRRs that comprise the 50 states of the U.S. (in addition, a breakdown for the USDA Forest Service's assessment regions established pursuant to the Resources Planning Act is

included in appendix 2). The WRRs are characterized by relatively homogenous precipitation, climate, geography, and water-use characteristics, although each unavoidably contains areas of considerable heterogeneity in some variables (figure 1).

The USGS improved its water-use data gathering procedure before preparing the 1985 circular. In addition to providing greater funding and more elaborate specifications to field offices collecting the data, the agency also changed some of the categories for which water use was summarized. Most of the changes resulted in more detail, much of it related to reporting about public-supplied water (involving delivery by a water supply entity such as a municipality or private water company serving multiple customers). Before 1985, deliveries from public supply to industrial and commercial users were grouped together, and deliveries from public supply to domestic and public uses were grouped together (the "public" in "domestic and public" refers to governmental office use, public parks, fire fighting, and losses in the public supply distribution system). Also, before 1985, mining and self-supplied commercial uses were grouped with self-supplied industrial. After 1985, public-supplied domestic, mining, self-supplied commercial, public-supplied commercial, self-supplied industrial, and public-supplied industrial were presented separately. Categories that were separate from 1960 through 1995 were self-supplied domestic (rural), livestock, self-supplied thermoelectric, and irrigation.

To obtain a small number of consistent categories for use in this report, USGS data were combined in two ways. First, self-supplies and public supplies were combined, as the source of supply was not an important distinction in this study. Second, the finer distinctions introduced in 1985 were not used, so that the categories were consistent for the entire 1960 to 1995 period. The following water-use[1] categories were chosen:

- livestock (self-supplied),
- domestic and public (public- and self-supplied),
- industrial and commercial (public- and self-supplied) and mining (self-supplied),
- thermoelectric power (self-supplied),
- irrigation (self-supplied), and
- hydroelectric power.

---

[1] In this report, the term water use is employed in a general way, to indicate any use of water, whether instream or offstream, and whether the offstream use is in terms of withdrawal or consumptive use. The USGS sometimes uses the term water use in a more specific way, to indicate the sum of self-supplied withdrawals and public-supplied deliveries. In this report, both self-supplied withdrawals and public-supplied deliveries are considered withdrawals. Also, when considering future years, the word use sometimes means quantity requested.

2

USDA Forest Service Gen. Tech. Rep. RMRS–GTR–39. 1999

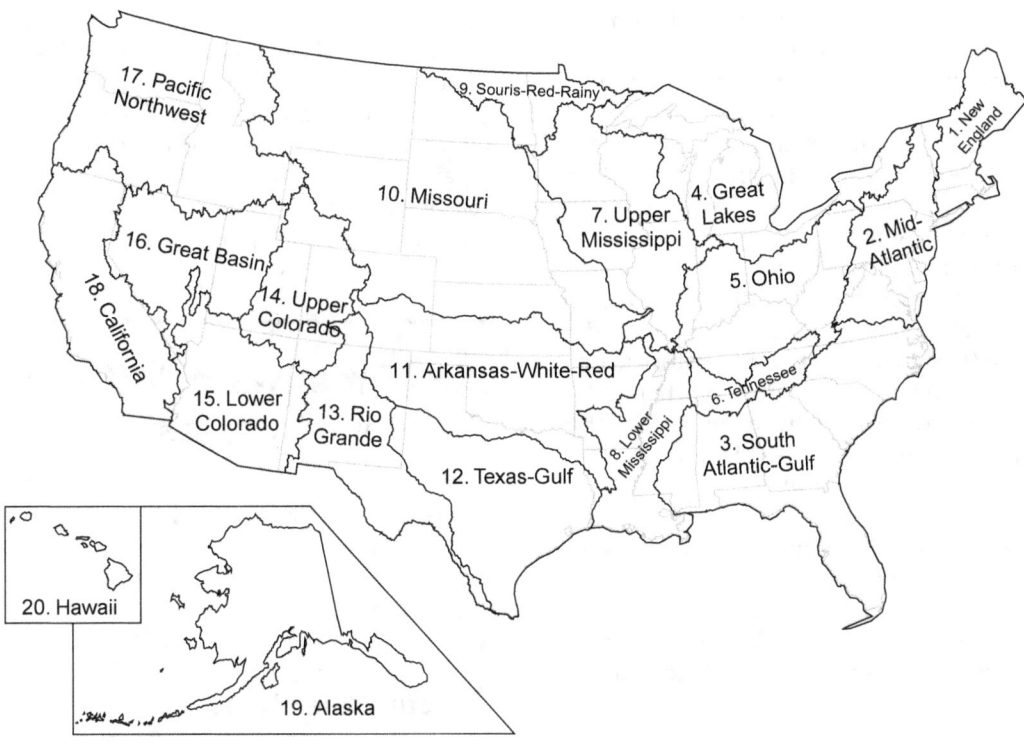

Figure 1. Water resource regions of the United States.

Except for hydroelectric water use, each of these categories can be expressed in terms of withdrawal or consumptive use.

Past water-use efficiency factors were computed using the USGS water-use data and data on water-use determinants. A ratio of the determinant to its respective quantity of water withdrawn (e.g., domestic withdrawal per capita) was computed for each use category. Projections of future levels of these water-use factors were made specifically for this study.

Total population was used as a determinant of future livestock, domestic and public, industrial and commercial, and thermoelectric water use. Historical population data were taken from the Bureau of the Census (1992) for the years 1960 to 1990 with the exception of the estimate for 1965, which was obtained from the Bureau of Economic Analysis (1992). Population projections through 2040 for the entire U.S. were obtained from the Census Bureau.[2] These projections were disaggregated to the state level using projected proportions from the Bureau of Economic

Analysis (1995). State-level projections were then disaggregated to the county level based on the distribution of state population to counties in 1990. County figures for past and projected future population were aggregated to WRRs using the county allocations of the Water Resources Council's Assessment Sub-areas (Water Resources Council 1978).[3] Past numbers of households, also investigated for estimating changes in domestic and public use, were taken from census records.

---

[2] Bureau of the Census, U.S. Department of Commerce, "Resident population projections of the United States: middle, low, and high series, 1996—2050," released March 1996, Washington, D.C.

---

[3] The USGS water-use circulars also list population by WRR. These estimates were not used, however, because investigation of trends showed a few large shifts from one time period to the next, suggesting that some of the estimates, especially for earlier years, may have been in error or that criteria for estimating population had changed. Because some of the USGS water withdrawal estimates for those years were based, at least partially, on the agency's population estimates, the water-use estimates for certain years may not correspond well with the Census Bureau population estimates reported herein. Thus, some estimates of per-capita water use shown in figures for specific WRRs, especially in earlier years, may be in error. To avoid related problems in projecting future water use, water-use efficiency factors involving population estimates were computed for the entire U.S.

Personal income was also used to project industrial and commercial water use. Historical data and projections for income per capita were obtained from the Bureau of Economic Analysis (1992 and 1993). As with population, county-level historical data and projections on income were aggregated to the WRR level using the county allocations of the Water Resources Council's Assessment Subareas (Water Resources Council 1978).

In addition to population, electricity production and assumptions about the distribution of that production among different types of generating plants were used to project water use at thermoelectric plants. Historical data on electric energy production for the entire U.S. from 1960, and by WRR since 1985, were taken from the USGS water-use circulars. Projections of future electricity production were made specifically for this study.

Number of irrigated acres and estimates of withdrawal per acre were used to project irrigation water use; historical data by WRR were obtained from the USGS circulars. Projections of future irrigated acres and withdrawal per acre were made specifically for this study. Table 1 lists the variables used to project water withdrawals for the five water-use categories.

The USGS water withdrawal estimates were sometimes based on assumed relations with other more easily measured variables, such as population or irrigated acres, rather than on actual measures of water diversion or delivery. The degree of reliance on assumed relations of withdrawal to other variables varied by water-use category, by USGS state office, and by year (with more recent estimates less likely to rely on assumptions). Any such reliance precludes independent efforts using the USGS data to discover what factors affected water use. Indeed,

only to the extent that the assumed relations were accurately specified can the USGS data be used to describe the relations of past water use to factors affecting that use and to project future water use. The limitations of the USGS water-use data, plus the difficulties of projecting future levels of each independent variable, are the principal reasons for using simple models when projecting water use for large geographical areas.

Past Freshwater Withdrawals in the United States

This section briefly describes water use for 1995, then depicts trends over the twentieth century, and finally looks in more detail at trends for 1960 to 1995, all for the U.S. as a whole.

Recent Water Use

USGS water-use data for 1995 indicate that, for the United States as a whole, hydroelectric plants used 3160 billion gallons per day (bgd), which is nine times the sum of all offstream withdrawals combined. The great majority of this use occurs instream (although not without disruptions to the aquatic environment). Looking at offstream use only, withdrawals (the sum of consumptive use and return flow/groundwater recharge in figure 2) totaled 340 bgd. The five categories of water use in figure 2 fall into three groups. The first group consists of the 2 largest users, agricultural irrigation and thermoelectric power, which each withdrew about 130 bgd. The second group consists of domestic and public use and industrial and commercial use, which each withdrew roughly 35 bgd. The fifth use, livestock, withdrew only about 5 bgd; however, much of U.S. irrigation is used to produce feed grains and forage

Table 1. Variables used to project freshwater withdrawals.

Water use category	Variable
Livestock	Population Withdrawal/person
Domestic & public	Population Withdrawal/person
Industrial & commercial	Population Income/person Withdrawal/dollar of income

Thermoelectric	Population
	Total kilowatt hours per person
	Freshwater thermoelectric
	  kilowatt hours/total kilowatt hours
	Freshwater thermoelectric
	  withdrawal/kilowatt hour

Irrigation	Acres irrigated Withdrawal/acre

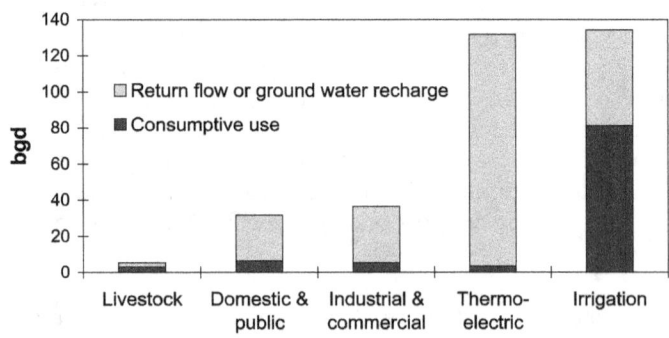

Figure 2. U.S. 1995 offstream water use.

for livestock. The water use that most citizens think of as water use, the domestic and public use category, accounts for only 9% of total freshwater withdrawal.

Consumptive use in 1995 totaled 100 bgd, or 29% of withdrawal (figure 2). Irrigation consumptively used 81 bgd. Consumptive use of the other four categories ranged from 3 bgd for livestock and for thermoelectric plants to 7 bgd for domestic and public use. Consumptive use is discussed in more detail in a later section.

Twenty-two percent of freshwater withdrawals in 1995 came from groundwater pumping; the remainder came from surface flows. Groundwater withdrawals are described in more detail in a later section.

Trends through the Twentieth Century

Growth in total U.S. water withdrawals during the twentieth century has, until recently, consistently outpaced population growth (figure 3). The changes in these 2 variables fall into 3 distinct periods. From 1900 to 1940, population increased by roughly 1.7 million persons per year while withdrawals increased by about 2.4 bgd per year. From 1950 to 1980, population increased by about 2.4 million persons per year while withdrawals increased by about 5.7 bgd per year. After 1980, total withdrawals dropped (and then leveled off, as seen below), but population continued to rise. Over the entire 1900 to 1990 period, population and withdrawal increased by 1.2% and 2.4% per year, respectively. Over the same period, total

withdrawals per capita increased by nearly a factor of four, from about 475 gallons per day in 1900 to about 1350 gallons per day in 1990 (figure 4).

The dramatic increase in withdrawals during this century is largely attributable to increases for irrigation and thermoelectric cooling, which together account for 83% of the total withdrawal increase from 1900 to 1990. Over this 90-year period, public supplies (domestic, commercial, and industrial) plus rural withdrawals (domestic and livestock) remained at roughly 12% of total withdrawals, self-supplied industrial withdrawals dropped from 25% to 6% of total, irrigation withdrawals dropped from 50% to 40% of total, and thermoelectric cooling withdrawals increased from 12% to 40% of total.

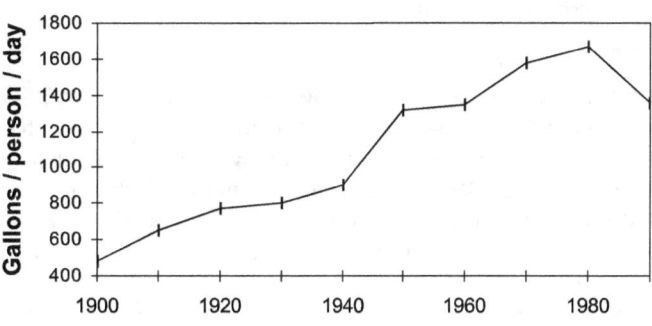

Figure 4. Total withdrawal per capita, 1900 to 1990.

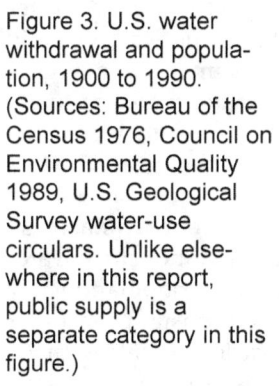

Figure 3. U.S. water withdrawal and population, 1900 to 1990. (Sources: Bureau of the Census 1976, Council on Environmental Quality 1989, U.S. Geological Survey water-use circulars. Unlike elsewhere in this report, public supply is a separate category in this figure.)

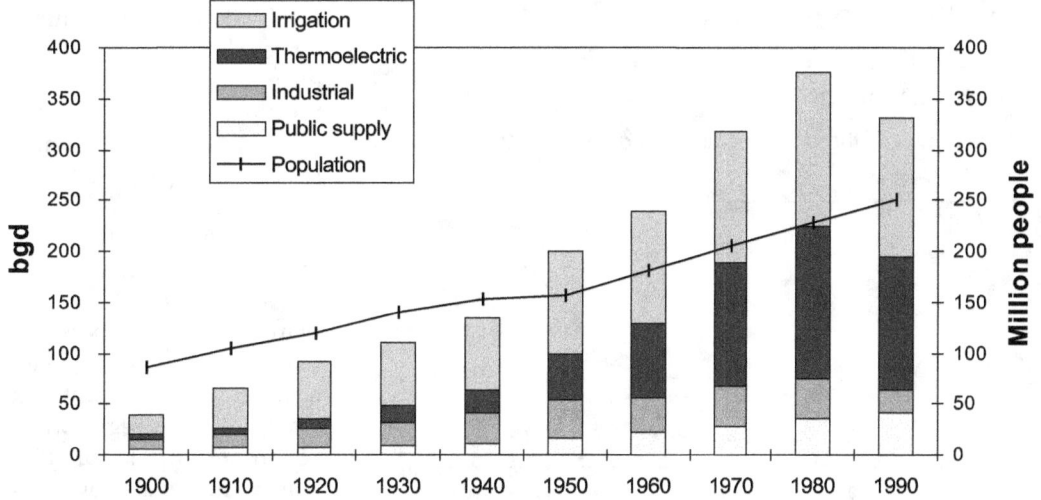

Trends Since 1960

Figure 3 shows a striking change in 1990, when total withdrawals dropped for the first time in the century. As that figure shows, the drop was not related to population trends. Figure 5, which presents withdrawals at five-year intervals since 1960 based on the USGS water-use circulars, shows that the drop first occurred in 1985, and that it is attributable to the top three categories: irrigation, thermoelectric, and industrial and commercial. Solley et al. (1988) suggested that the drop in 1985 was due partly to

- above average rainfall that year, which lessened the need for irrigation withdrawals,

- an economic slowdown and lower commodity prices,

- higher groundwater pumping costs as lifts had continued to increase, and

- improved efficiency in water use.

Solley et al. also suggested, however, that the drop in 1985 was partially attributable to the improved process for amassing the water-use data that was initiated by the USGS for the 1985 report, and concluded that earlier estimates may have been too high. A further factor, most important for irrigation, is the subsidence of the era of large-scale, federally funded water developments. Dam construction continues to increase the available water supply, but the rate of change has greatly diminished after peaking during the 1960s (figure 6).

The fact that the three major water uses changed little between 1985 and 1995, although rainfall in 1990 was below that in 1985 and economic conditions improved, suggests that rainfall and general economic conditions did not play deciding roles in the dramatic 1985 drop in withdrawals. Although higher pumping costs, improving efficiency of water use, and the waning of the dam construction era undoubtedly contributed to the 1985 drop, such factors have had a gradual and continuing influence over many years and thus, are unlikely to have been wholly responsible for the abrupt 1985 drop. The change in the USGS's estimation procedure appears to have played a significant role in the reported 1985 drop in withdrawals. This possibility highlights the importance of focusing on long-term trends rather than short-term shifts when using the USGS water-use data.

The following subsections discuss trends in withdrawals for each of the five water-use categories shown in figure 5, beginning with the smallest use category, livestock, and ending with the largest, irrigation. Next, trends in hydroelectric water use are briefly described. Tables A1.1 to A1.5 list withdrawals for the five water-use categories by decade beginning with 1960.

Livestock

The USGS's livestock water-use category consisted of use by terrestrial animals (called stock, principally cattle, hogs, sheep, and poultry) until 1985, when animal specialties (principally fish farming) were moved from the industrial to the livestock category. Once the ponds are established, water is needed at fish farms to maintain pond levels.

Use by terrestrial animals was estimated by the USGS largely based on numbers of animals served, with different animal species assigned their respective average water requirements. Use of water at fish farms was typically estimated based on pond area and estimates of evaporation and seepage. U.S. livestock withdrawals gradually increased from 1960 to 1980 in response to increasing animal numbers, then more than doubled in 1985 when animal specialties were added (bars in figure 7).

Estimates of future stock numbers are unavailable, so using animal numbers as the determinant of stock water use was not promising. However, human population may serve as a determinant because, given constant consumer tastes, meat and egg consumption varies roughly with the consumer population. Figure 8 shows total and per-person withdrawals for stock. Withdrawal per person has remained quite constant over the past 35 years, ranging between 9 to 10 gallons per day (dots in figure 8).

Adding water use for animal specialties complicates the picture, raising withdrawal per person to from 18 to 21

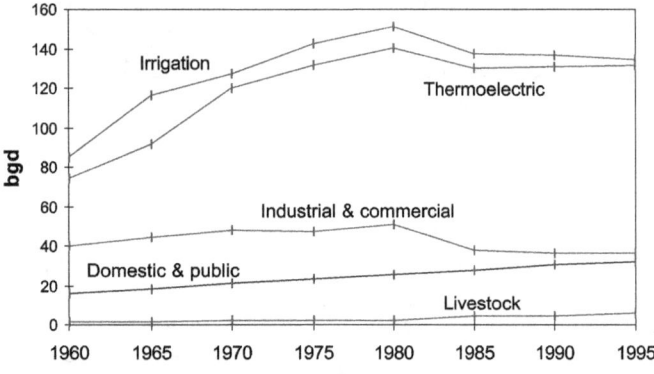

Figure 5. U.S. withdrawals by use category, 1960 to 1995.

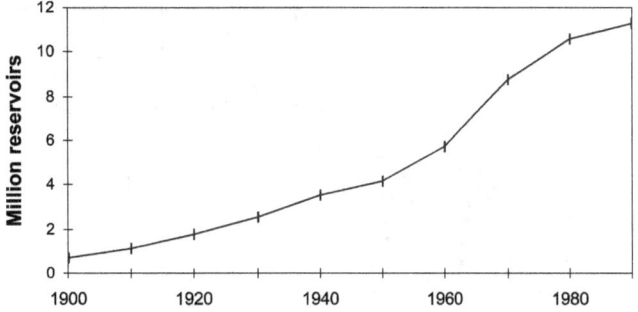

Figure 6. Cumulative number of reservoirs in the U.S., 1900 to 1990. (Source: Army Corps of Engineers and Federal Emergency Management Agency 1992)

6

USDA Forest Service Gen. Tech. Rep. RMRS–GTR–39. 1999

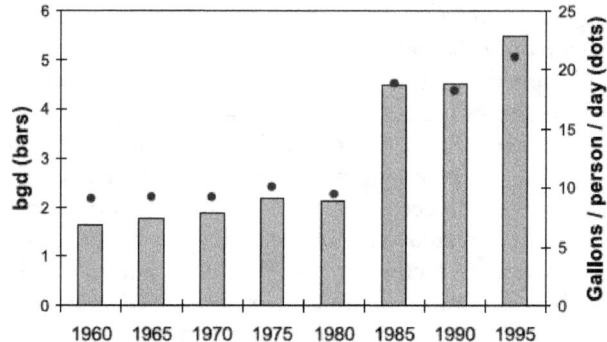

Figure 7. Livestock withdrawal in the U.S.

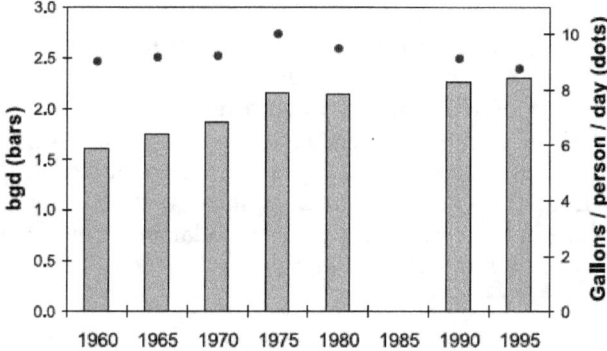

Figure 8. Stock withdrawal in the U.S. (Withdrawals for stock and for animal specialties were not separated in the 1985 USGS water use circular.)

gallons per day since 1985 (dots in figure 7). The 3-gallon per person per day change from 1990 to 1995 is attributable to increased water use by animal specialties, suggesting that per-capita water use in aquaculture may be growing.

Domestic and public use

Total U.S. withdrawals for domestic and public water use (public uses and losses are about 15% of total domestic and public withdrawals) consistently increased during the 1960 to 1995 period, rising from 16 bgd in 1960 to 32 bgd in 1995 (figure 5). The continued rise after 1985 contrasted with the other three major uses, which all reversed their prior upward trend (figure 5). One possible reason for the continued rise in domestic and public withdrawal is that these uses are relatively unresponsive to price.[4] Another reason may be that the USGS estimates of domestic and

public water use have been partially based on population estimates.

The rise in domestic and public withdrawal was primarily caused by population growth, but population increase is not the whole story. As seen in the light bars of figure 9, per-capita domestic and public withdrawal also steadily increased, from 89 gallons per day in 1960 to 122 in 1990. This increasing per-capita water use may be largely attributable to a decrease in average household size (Schefter 1990). As shown by the dots in figure 9, people per household (i.e., per occupied housing unit) decreased from about 3.4 in 1960 to 2.7 in 1990.[5] A minimum level of water use per household, especially for lawn and garden watering, is largely unrelated to household size, causing per-capita use to rise as household size drops.

Other factors probably contributing to the increase in per-capita domestic water use include the conversion of older or rural households to complete plumbing, and an increase in the use of appliances such as dishwashers, washing machines, swimming pools, and lawn sprinkler systems. These changes are consistent with the increasing real incomes experienced in many areas of the U.S. over the past 30 years.[6]

The consistent growth in per-capita domestic and public withdrawals since at least 1960 may have ended, as per-capita withdrawals dropped from 122 gallons per day in 1990 to 120 in 1995 (figure 9). This change may be the result of several factors. First, the drop in number of persons per household may have ended; it dropped by 1.3% per year in the 1970s, 0.4% per year in the 1980s, and by only 0.03% per year from 1990 to 1995 (dots in figure 9). Second, the conversion of older or rural houses to modern plumbing, another cause of the previous rise in per-capita withdrawal, is nearly complete. Third, many public suppliers have begun encouraging conservation by:

- adding meters to unmetered houses,

- educating customers about conservation, and

- altering pricing structures to discourage excess use.

---

[4] Numerous studies have shown that demand for domestic water is relatively inelastic to price. This inelasticity may occur because domestic uses are highly valued, to some extent essential, and require little of a household's or municipality's budget. See, for example, Williams and Suh (1986), Diaz and Brown (1997, chapter 2), or Espey and others (1997).

[5] The decrease in household size may be due to various demographic trends such as increasing longevity, decreasing fertility rate, and increasing divorce rate.

[6] These changes are also consistent with decreasing real domestic water prices. Supporting this notion, Schefter (1990) reported a drop in the average real price for domestic water across the U.S. from 1968 to 1984. However, he was unable to include sewer prices, which may have been rising at the same time as utilities worked to establish separate wastewater pricing schemes. Thus, it is unclear to what extent changing water prices caused the increase in withdrawal per capita.

Fourth, new plumbing fixture standards, promulgated in the Energy Policy Act of 1992, have taken effect.[7] Despite these reasons for a trend reversal, it is too soon to be sure whether the recent change in per-capita use trend will persist.

To the extent that the increase in domestic withdrawal per capita was caused by conversion to complete plumbing and use of more water-using appliances, we would expect withdrawal per household to also have risen. However, although withdrawal per household grew from about 300 gallons per day in 1960 to 326 in 1995, estimates for 1970 and 1990 are slightly higher than the 1995 estimate (dark bars in figure 9). The overall 35-year record suggests an increase, but the 1970 estimate, especially, sheds doubt on the trend.[8] Because of the confusing data on withdrawal per household, it was decided to not use number of households in the projection of domestic and public use and thus, to focus only on population and use per capita.

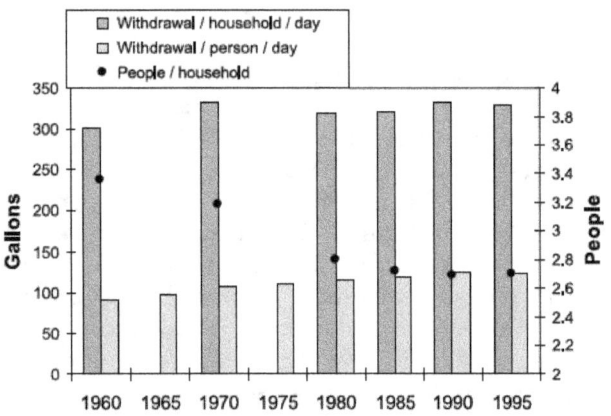

Figure 9. Domestic and public withdrawal in the U.S.

Industrial and commercial use

Total U.S. industrial and commercial withdrawals show a gradual rise from 1960 to 1980, then the sharp decrease in 1985 discussed earlier (figure 5). Only about 2.3 of the 13 bgd drop from 1980 to 1985 is attributable to moving animal specialties to the livestock water-use category. Since 1985, total withdrawals have remained at about 36 bgd.

---

[7] Section 123 of the Energy Policy Act of 1992 (Public Law 102-486) set standards for the "maximum water use allowed" for certain types of fixtures manufactured after January 1, 1994. For example, lavatory faucets were restricted to 2.5 gallons per minute at a water pressure 80 pounds per square inch, and gravity tank-type toilets were restricted to 1.6 gallons per flush.

[8] The 1970 data point is open to question; examination of trends in number of households versus population shows that the 1970 estimate of number of households is questionably low.

Although industrial and commercial withdrawals increased from 1960 to 1980, withdrawals per unit of output in all major industrial sectors decreased during that period (David 1990). Because of the great variety of outputs of the industrial and commercial sectors, relating water use to units of physical output was unrealistic for this study. Instead, an economic measure of total output, personal income, was used. Withdrawals per dollar of total personal income declined steadily, from 24 gallons in 1960 to 7 gallons in 1995 (bars in figure 10). The drop in withdrawal per dollar of income is largely attributable to changes in the type and quantity of industrial and commercial outputs, such as a shift from water intensive manufacturing and other heavy industry to service oriented businesses, and to enhanced efficiency of water use. Efficiency has improved in response to factors such as environmental pollution legislation (e.g., the Clean Water Act of 1972 and its amendments), which regulated discharges and thereby encouraged reductions in withdrawals, and technological advances facilitating recycling.[9] The most recent data show that the rate of decrease in water withdrawal per dollar of income has slackened somewhat (bars in figure 10).

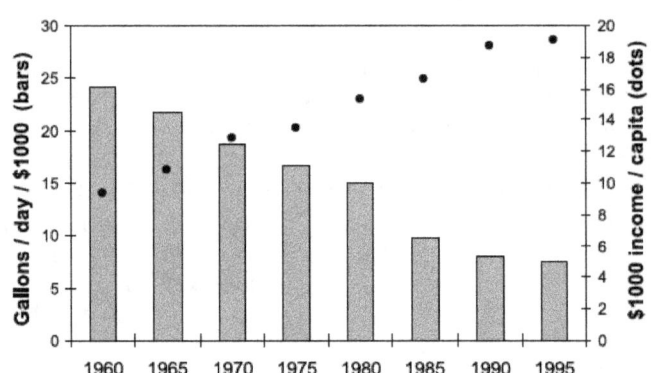

Figure 10. Industrial and commercial withdrawal in the U.S. per dollar of income. (Per-capita income in 1990 dollars.)

Thermoelectric use

At thermoelectric power plants (mainly fossil fuel and nuclear plants), water is used principally for condenser and reactor cooling. Freshwater withdrawals increased steadily through 1980, declined substantially in 1985, as mentioned, and have increased only slightly since then (figure 5). Withdrawals of saline water, not shown in

---

[9] As David (1990) points out, the environmental pollution legislation essentially raised the cost of withdrawing water to industrial users. In response to this price rise, industries lowered withdrawals per unit of output by modifying production processes and increasing recycling of withdrawn water.

figure 5, have equaled roughly 30% of total water withdrawals at thermoelectric plants since 1960.

In contrast to the recent leveling off of total withdrawals, production of electricity at freshwater thermoelectric plants has continued to rise (dots in figure 11). Indeed, freshwater withdrawal per kilowatt hour (kWh) produced has declined steadily, and in 1995 it was only 42% of its 1960 value (bars figure 11). This improvement in the efficiency of withdrawals has allowed thermoelectric energy production using freshwater to increase by 322% for a mere 78% increase in withdrawal. The improved efficiency has occurred partly by greater reuse of withdrawn water; during the 35-year period, consumptive use by thermoelectric plants increased by a factor of 14 (although, as seen in figure 2, consumptive use is still a small fraction of withdrawal). The latest data indicate a leveling off of the rate of decrease in withdrawal per kilowatt hour (bars in figure 11).

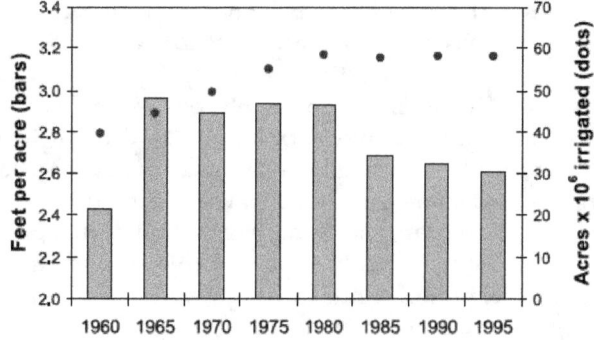

Figure 12. Irrigation withdrawals per irrigated acre in the U.S.

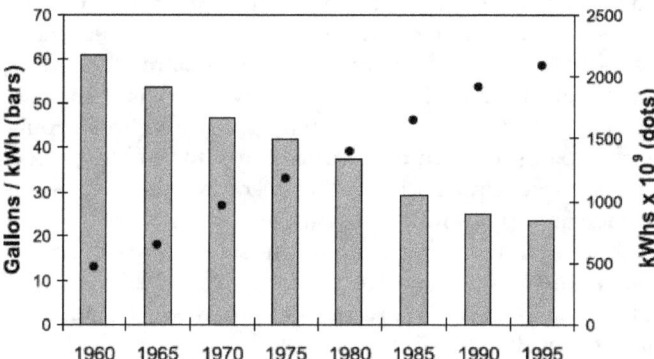

Figure 11. Thermoelectric freshwater withdrawal per kilowatt hour in the U.S.

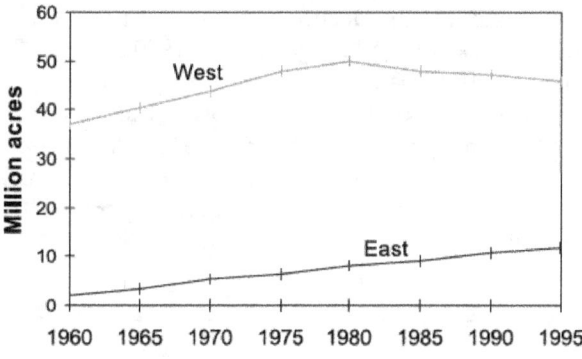

Figure 13. Irrigated acreage in the U.S.

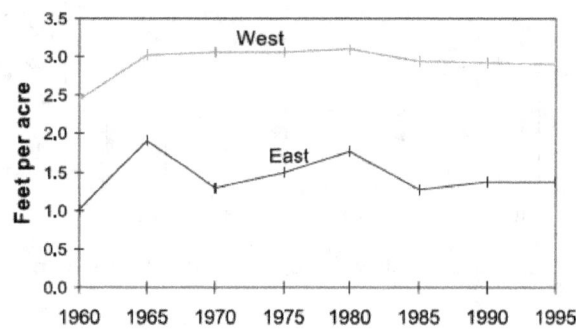

Figure 14. Depth of irrigation withdrawal in the U.S.

Irrigation

U.S. withdrawals for irrigation steadily increased from 1960 to 1980, then declined in 1985, with additional smaller decreases since then (figure 5). The decreases since 1985 are not a simple function of irrigated acreage changes, as overall irrigated acreage rose from 57.2 million acres in 1985 to 57.9 million acres in 1995 (dots in figure 12). Instead, a geographical shift in irrigated acreage occurred. The arid and semi-arid West, where the vast majority of irrigation occurs, is experiencing a decrease in irrigated acreage that began in the early 1980s, as farmers sell some of their land or water to cities, industries, and rural domestic users, or as pumping costs cause marginal lands to be removed from irrigation. At the same time, farmers in the East are relying more on irrigation water to supplement precipitation during dry times, to reduce variability in yields and product quality (Moore and others 1990). This phenomenon is depicted in figure 13, where the East is WRRs 1 through 9 and the West is WRRs 10 through 18.

The drop in irrigated acreage in Western regions, which tend to use relatively large amounts of water per acre, and rise in irrigated acreage in Eastern regions, which use relatively less water per acre, is partly responsible for the nationwide drop in water application per acre that began in 1985 (bars in figure 12).

The recent downward trend in withdrawal per acre (figure 12) is also attributable to a decrease in per-acre water applications. Application rates dropped in the East and West from 1980 to 1985, and they have continued to drop in the West (figure 14). The portion of withdrawal that is consumptively used is one indication of irrigation efficiency; improved methods withdraw less water for a given amount of plant transpiration. From 1985 to 1995, consumptive use increased from 47 to 59% of withdrawal in the West. If these estimates are accurate (note that measures of consumptive use rely on a good deal of educated judgment), they corroborate the drop in withdrawal per acre. Improved irrigation efficiency may be a response to factors such as the waning of the era of publicly-funded dam and canal construction, higher prices for water from publicly-funded projects, increasing groundwater pumping lifts, and improved irrigation technology (Moore and others 1990).

Hydroelectric use

Water use for hydroelectric energy generation increased from about 2×10^{12} gallons per day in 1960 to about 3.3×10^{12} in 1975, but this category of water use has remained rather constant since then. Total kilowatt hours produced shows the same pattern (dots in figure 15). Water use per kilowatt-hour produced has remained roughly constant since 1965, averaging 4.1 thousand gallons per kWh (bars in figure 15).

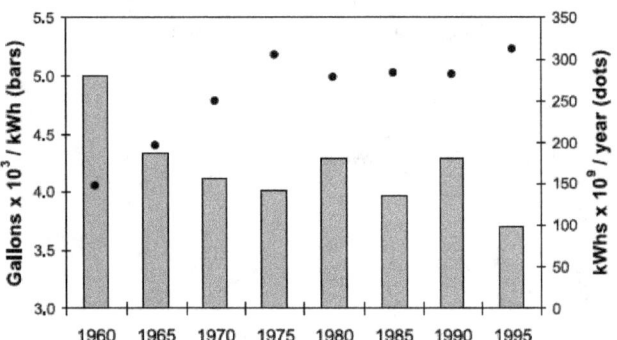

Figure 15. Hydroelectric water use per kilowatt hour in the U.S.

Projection of Freshwater Withdrawals

The following projections of withdrawals are based on estimates of future population and income provided by the Bureau of the Census and Bureau of Economic Analysis and on explicit assumptions about rates of change in other factors affecting water use developed specifically for this study. In some cases (e.g., industrial and commercial withdrawal per dollar of income), these future rates of change extend consistent past trends. In other cases (e.g., domestic and public use per capita), recent abrupt changes in past trends have made trend extrapolation problematic.

In either situation, assumed rates of future change in other factors affecting water use do not reflect a detailed model, economic or otherwise. Rather, they were chosen to maintain the visual continuity of the trend, as will be apparent in subsequent figures, or to reflect conjecture in light of recent trend shifts. Extension of past trends is justified on the assumption of continuation of the fundamental forces affecting past changes.

Extending past trends in other factors affecting water use usually required a diminishing rate of change. Computationally, estimates of future levels of these factors were specified by applying an annual rate of change (i) to the quantity (Q) of the prior time period. Quantity for year n was computed as: $Q_n = Q_{n-t} \cdot (1 + i_n)^t$, where t is time period in years and $i_n = i_{n-t} \cdot (1 + d)^t$, where d is an annual change in i chosen to maintain continuity of the prior trend. When n = 2000, $t = 5$, Q_{n-t} is the estimate for 1995, and i_{n-t} is the recent historical trend.[10] When n = 2010, 2020, 2030, or 2040, $t = 10$, and Q_{n-t} and i_{n-t} are the projected quantity and rate, respectively, of the prior time period. Rates i and d were selected separately by water-use factor. Rate of change (i) was positive, negative, or nil depending on the prior trend. Except in cases where i is 0 and d is immaterial, decay (d) was always negative, in keeping with the observation of diminishing rate of change (see, for example, figures 10 and 11). The results of this approach are apparent in subsequent figures.

Annual rates of change (i) for most uses were specified only at the national level, considering that the fundamental forces affecting future rates of change in withdrawal are not localized.[11] Thus, rates of change (i) for livestock and domestic and public withdrawals per person, industrial and commercial withdrawal per dollar of income, thermoelectric withdrawal per freshwater kilowatt hour, and total kilowatt hours per person were specified for the entire U.S. These rates were then applied at the WRR level to 1995 estimates of Q_{n-t}. Agricultural irrigation was considered subject to more region-specific forces than the other factors. Thus, as seen below, rates of change (i) for acres irrigated were specified at the WRR level, and rates

---

[10] The following cases are exceptions to setting Q_{n-t} for n = 2000 equal to the 1995 value: 1) when the 1995 estimate showed a distinct shift from 1990, Q_{n-t} was usually set equal to the mean of the estimates for years 1990 and 1995; 2) for irrigation withdrawal per acre, the 1985 to 1995 mean was used to lessen the temporal effects of weather, except in Alaska where, because of missing data, only the 1990 and 1995 rates were averaged.

[11] Also, the withdrawal estimates reported by the USGS at the WRR level, especially for earlier years, exhibit some unusual shifts. Because so much time has passed and personnel have changed, explanations for some of these shifts are difficult to obtain. These shifts complicate analysis of trends of water-use factors such as domestic withdrawals per capita. Using aggregate U.S. data alleviates such data problems.

for withdrawal per acre were specified separately for the Eastern and Western portions of the U.S.

Projection of National Withdrawals

This section describes water-use projections and the assumptions behind those projections for the five water-use categories, beginning with the smallest withdrawal category. Figure 16.1, .2, .3 depicts projections for all categories at the national level. In this and subsequent figures, dark bars indicate past withdrawals and light bars indicate future withdrawals; similarly, dark dots show past levels of related factors and light dots show future levels.

As explained, population is a variable in the projections of four of the five withdrawal categories. Population growth has gradually lessened in percentage terms, from 1.3% per year in the 1960s, to 1.1% in the 1970s, to 0.9% in the 1980s. The Census Bureau's middle series projections (figure 16.1) show U.S. population increasing at annual rates of 1% during the 1990s, 0.8% from 2000 to 2010, and about 0.7% thereafter. This consistent growth rate is apparent in figure 17 and in the top 2 graphs of figure 16.1, which show total U.S. population increasing along a nearly straight line, rising from 263 million in 1995 to 370 million in 2040, a 41% increase.

The population projections are based on assumptions about three factors: life expectancy, fertility, and immigration. The Census Bureau estimates high, medium, and low levels for each factor. The middle series projections use the three medium-level assumptions, which are an average life expectancy of 82 years, an average fertility rate of 2.245 births per woman, and a net annual immigration of 820,000 persons.[12] Likewise, the high series projections use the high-level assumptions for each factor, and the low series projections use the low level assumptions for each factor. Mixing, for example, low and medium level assumptions for the three factors can produce intermediate projections (not used herein). The Census Bureau does not present confidence limits about these different estimates. As seen in figure 17, by the year 2040 there are considerable differences among the series. The low and high series show changes in total population from 1995 to 2040 of 9% and 74%, respectively, in comparison with the 41% increase with the middle series.

The following sections on the five water-use categories report results based on the middle population series (figure 16.1). For comparison with figure 16.1, estimates using other population series are presented in figures 16.2 and 16.3, with accompanying text at the end of this section.

---

[12] For details, see the 1996 Census Bureau release in footnote 2.

Middle population series

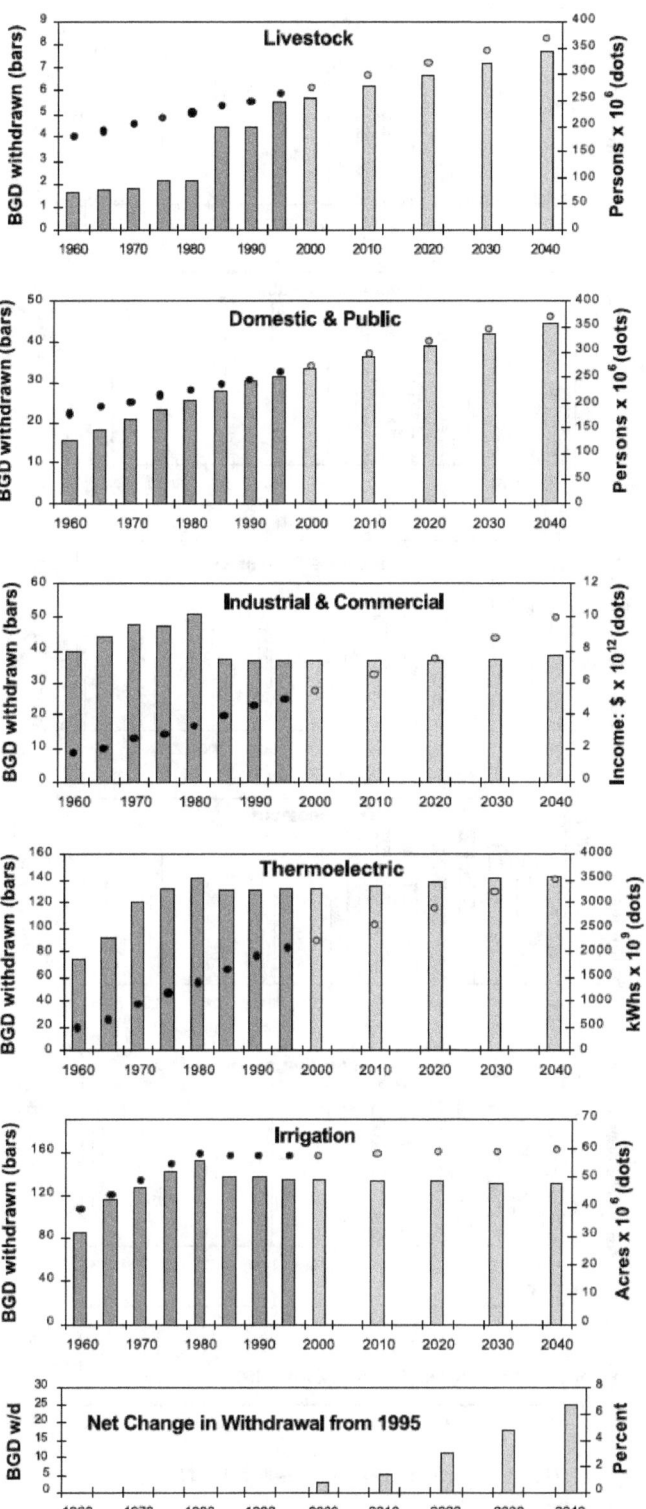

Figure 16.1. Withdrawal projections for the U.S. using middle population projection.

Low population series

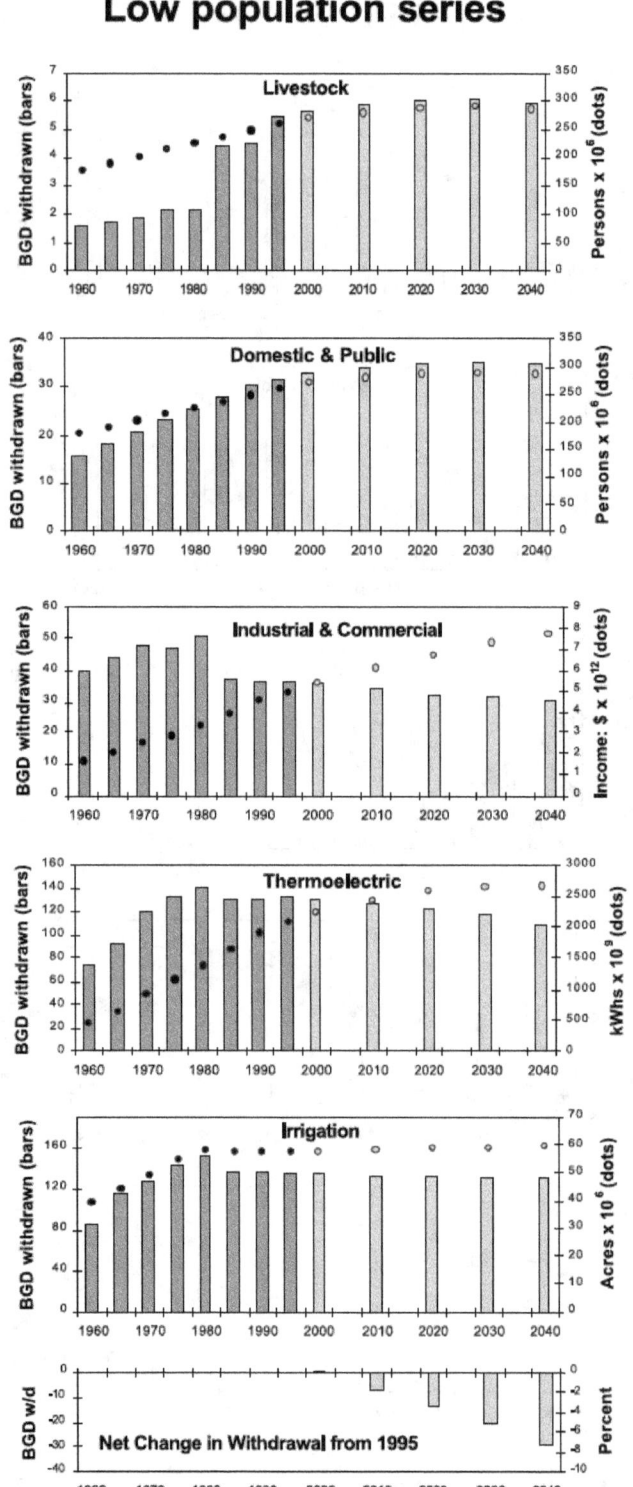

Figure 16.2. Withdrawal projections for the U.S. using low population projection.

High population series

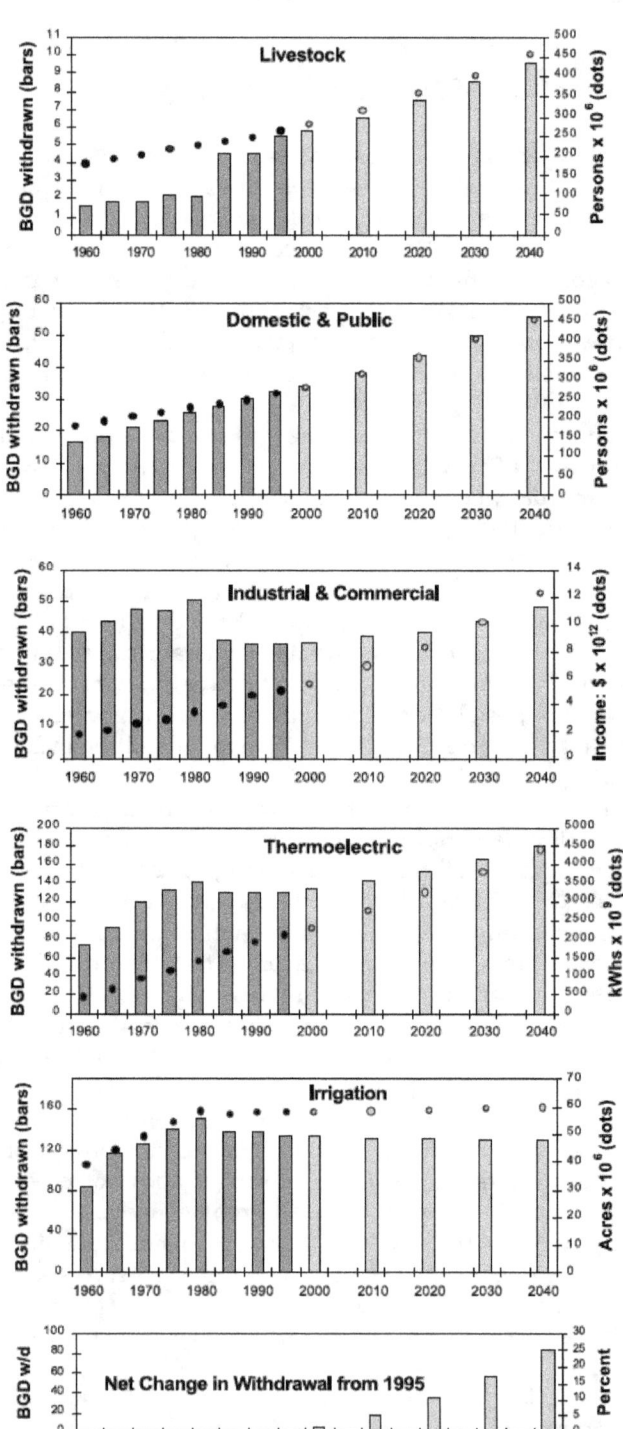

Figure 16.3. Withdrawal projections for the U.S. using high population projection.

12

USDA Forest Service Gen. Tech. Rep. RMRS–GTR–39. 1999

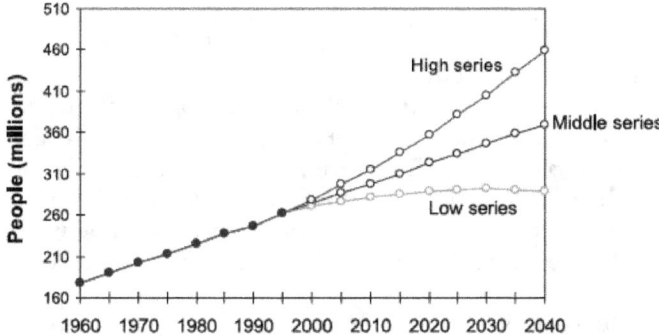

Figure 17. Population projections for the U.S.

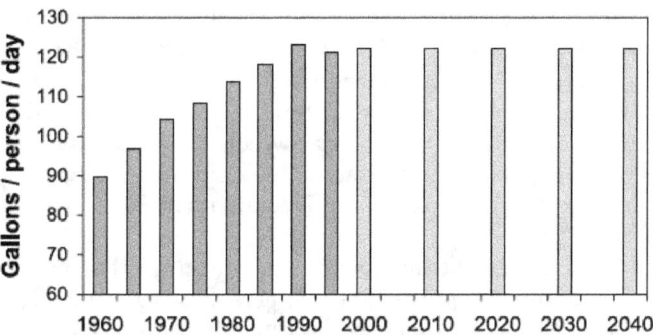

Figure 18. Projected domestic and public per-capita withdrawal.

Livestock use

Livestock withdrawals per person were assumed to remain constant at the 1995 level of about 21 gallons per day. This assumption allows for shifting human tastes within the livestock category, consisting, as mentioned earlier, largely of beef, pork, lamb, chicken, eggs, and farm-grown fish. The assumption ignores the implication of the 1990 to 1995 increase in withdrawal per capita for aquaculture (figure 7), which is that additional increases may be in store, because in this case a single year's change is assumed to be insufficient to indicate a trend. Total livestock withdrawal in the U.S. is projected to rise from 5.5 bgd in 1995 to 7.7 bgd in 2040 (figure 16.1).

Domestic and public use

Domestic and public withdrawals were projected based on population and per-capita withdrawal. Specifically, domestic and public withdrawals were projected as: population · (withdrawal / person).

After consistently increasing for at least 30 years at annual rates of 1.5% during the 1960s, 0.9% during the 1970s, and 0.8% during the 1980s, withdrawal per person dropped by 0.3% per year from 1990 to 1995. This change might be ignored as too recent and too small to indicate a major shift in the prior trend. However, several factors listed above (the end of the drop in household size, the completion of conversion to modern plumbing, and the growing impact of conservation measures) suggest that a significant trend change may be occurring.

It is impossible to say what will happen to per-capita domestic and public withdrawals in the future. Trends up to 1990 suggest continued growth, but recent changes suggest future decreases. Given this conundrum, it is assumed here that future per-capita withdrawal will remain constant at 121 gallons per day, equal to the midpoint between the 1990 and 1995 levels (figure 18). National domestic and public withdrawals are thus projected to increase at the same rate as population, from 32 bgd in 1995 to 45 in 2040, a 42% increase,

assuming the middle population protection series (figure 16.1).[13] Alternative assumptions about per-capita withdrawal are examined in the sensitivity analysis section.

Industrial and commercial use

Industrial and commercial withdrawals were projected based on estimates of future population and income and assumptions about the rate of change in withdrawal per dollar of income. Specifically, withdrawals were projected as: population · (dollars of income / capita) · (withdrawal / dollar of income).

The Bureau of Economic Analysis projects per capita income, in 1990 dollars, to increase from $19,001 in 1995 to $27,103 in 2040 (dots in figure 19), which is equivalent to a growth rate of about 0.8% per year. Withdrawal per dollar of income, which dropped at annual rates of 2.5% during the 1960s, 2.3% during the 1970s, and 6.1% during the 1980s, but by only 1% from 1990 to 1995, was assumed

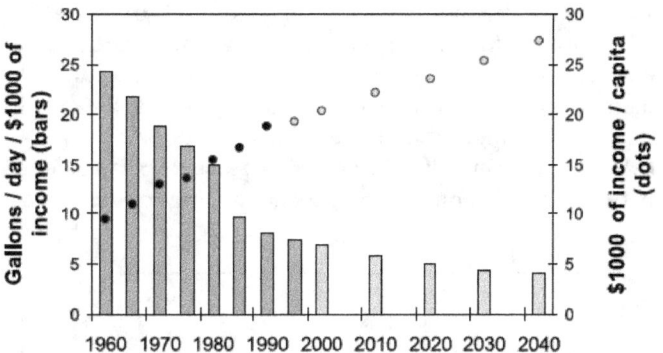

Figure 19. Projected industrial and commercial withdrawal factors.

---

[13] *This percentage increase is slightly higher than the 41% increase in population because the constant rate of future per-capita withdrawal was, in accordance with the method followed when a recent trend shift occurred, set equal at the midpoint between the 1990 and 1995 levels, 121 gallons per day, rather than at the 1995 level of 120 gallons per day.*

USDA Forest Service Gen. Tech. Rep. RMRS–GTR–39. 1999

13

to drop at a gradually decreasing rate of from 2% to 1% per year from 1995 to 2040. This assumption continues the past trend of conversion to more efficient processes and greater levels of water recycling. Given this assumption, withdrawal per $1000 of income, which dropped from 24 gallons in 1960 to 7.4 gallons in 1995, continues dropping but at a lower rate, reaching 3.9 gallons in 2040 (bars in figure 19).

Total industrial and commercial withdrawals are projected to remain quite stable, rising from 37 bgd in 1995 to 39 in 2040, a 5% increase, assuming the middle series population projections (figure 16.1). Thus, the decreasing withdrawal per dollar largely compensates for the continued increases in population and per-capita income. On a per-capita basis, industrial and commercial withdrawals are projected to decrease from 140 gallons per day in 1995 to 106 in 2040 (per-capita withdrawals by decade are listed in table A1.7).

Thermoelectric power use

Withdrawals at thermoelectric plants were projected based on estimates of future population and assumptions about the rate of change in energy use per person and in water use per kilowatt hour produced, plus an assumption about the proportion of total energy production that will occur at freshwater thermoelectric plants. Specifically, freshwater withdrawal for electricity production was estimated as: population · (total kWhs / person) · (freshwater thermoelectric kWhs / total kWhs) · (freshwater withdrawal / freshwater thermoelectric kWh).

Total energy use (thermoelectric plus hydroelectric) per person rose from about 4200 kWh per year in 1960 to about 11,400 in 1995 (dots in figure 20); this rise proceeded at annual rates of 6% during the 1960s, 3% during the 1970s, 1.1% during the 1980s, but only 0.4% from 1990 to 1995. In keeping with this decreasing trend, future total energy use was assumed to increase by an annual rate decreasing from 0.6% to 0.14% from 1995 to 2040 (dots in figure 20), bringing total energy use per person to 13,040 kWhs per year in 2040. This rate of increase reflects a balance between development of more energy using conveniences,

which would lead to greater energy use per person, and improvements in energy efficiency of all such conveniences, which would lead to less energy use per person.

Further, it was assumed that generation at hydroelectric plants remained constant at the 1995 level (it has been quite stable since 1975, figure 15) so that all increases in production occurred at thermoelectric plants, and that the allocation of thermoelectric energy production between freshwater and saltwater plants remained constant at the 1995 level.[14] Given these two assumptions, which were applied at the WRR level, and given the projected increase in total electric energy consumption described above, annual use of energy generated at freshwater thermoelectric plants, which increased from 2493 kWhs per person in 1960 to 7917 in 1995, was assumed to reach 9,421 kWhs per person in 2040. This trend, along with the expected population increase, produces the increases in total annual energy production at freshwater thermoelectric plants depicted in figure 16.1 (from $2.1 \cdot 10^{12}$ kWhs in 1995 to 3.5 $\cdot 10^{12}$ in 2040).

Freshwater use per kilowatt hour produced at thermoelectric plants decreased at annual rates of 2.7% from 1960 to 1985 and 2.0% from 1985 to 1995. In keeping with this apparent leveling off of the rate of decrease, future water use per kilowatt hour was assumed to decrease by from 1.3% to 0.6% per year from 1995 to 2040. Given this rate of decrease, water use per kilowatt hour produced at freshwater thermoelectric plants, which decreased from 60 gallons per kWh in 1960 to 23 in 1995, reaches 16 gallons per kWh in 2040 (bars in figure 20). This trend, along with the increase in electricity production, causes total freshwater withdrawal to rise from 132 bgd in 1995 to 143 in 2040, a 9% increase, assuming the middle series population projections (figure 16.1). Thus, the decreasing withdrawal per kilowatt hour is projected to only partially compensate for the increases in electricity production required to accommodate the growing population and per-capita energy use. On a per-capita basis, thermoelectric freshwater withdrawals are projected to decrease from 504 gallons per day in 1995 to 389 in 2040.

Irrigation

Many factors affect agricultural irrigation withdrawals. Irrigation is a lower-valued use of water at the margin than most other uses, so that withdrawals for irrigation in water-short areas are partially a function of water use in

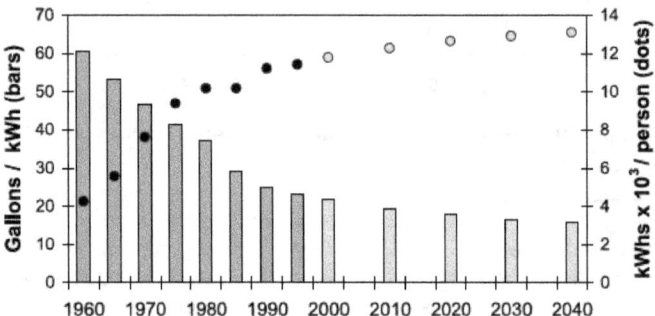

Figure 20. Projected electric energy withdrawal factors.

---

[14] In 1995, about 10% of electric energy production in the U.S. occurred at hydroelectric plants (this percentage varied from 1% in WRRs 5 and 7 to 89% in WRR 17), and about 20% occurred at saline water thermoelectric plants (this percentage varied from 0% for most inland WRRs to 91% in WRR 20). The remaining electric energy production (70%) occurred at fresh water thermoelectric (this percentage varied from 1% in WRR 18 to 99% in WRRs 5 and 7).

more highly valued uses.[15] In addition, irrigation water use is a complicated function of population, as population growth both increases demand for crops and, via urban expansion, decreases availability of irrigable land. Other factors affecting irrigation include energy prices (especially their effect on pumping costs), irrigation technologies, international markets for agricultural crops, changing tastes for livestock (nearly half of Western irrigated land is used to produce feed and forage for livestock), federal agricultural policies, instream flow concerns, and precipitation variations. Because accounting for all these factors is problematic, a simple approach was adopted for estimating future irrigation withdrawals that sets withdrawal equal to irrigated acreage · (withdrawal / acre). Future acreage and withdrawal per acre were estimated by extrapolating past trends.

Because some of the factors affecting irrigated acreage vary considerably by region, such as availability of irrigable land and competition for water, acreage changes were estimated at the WRR level, as described in the section on WRR projections. This approach yields a projection of U.S. irrigated acreage increasing gradually from 57.9 million acres in 1995 to 62.4 million acres in 2040 (figure 16.1).

Withdrawal per acre varies considerably from year to year at the WRR level because of weather. Thus, time trends of withdrawal per acre at the WRR level are often erratic. To avoid this localized phenomenon, withdrawal per acre was investigated for 2 large regions: the East and the West. As seen in figure 14, the decrease in withdrawal per acre in the West has been consistent from 1980 to 1995, whereas in the East there has been no consistent trend since 1985. In the West, withdrawal per acre, which fell at annual rates of 1% from 1980 to 1985 and 0.1% from 1985 to 1995, was assumed to continue falling at a rate of from 0.08% to 0.04% per year from 1995 to 2040. Given these rates, withdrawal per acre in the West, which dropped from 3.10 feet in 1980 to 2.91 feet in 1995, drops to 2.84 feet by 2040 (figure 21). In the East, withdrawal per acre was assumed to remain constant. These rates of decrease in withdrawal per acre, 0% per year in the East and 0.08% to 0.04% per year in the West, were applied to a beginning rate in each WRR set equal to the mean for the years 1985, 1990, and 1995. The overall drop in withdrawal per acre in the West from this 1985 through 1995 mean to the year 2040 is 2.9%.

The aggregation of the results of applying the estimates of acreage and withdrawal per acre at the WRR level yields a slightly decreasing level of total withdrawal for the U.S., dropping from 134 bgd in 1995 to 130 in 2040 (figure 16.1). On a per-capita basis, irrigation withdrawals are projected

---

[15] One indication of the relatively low marginal value of agricultural water is that most of the recent water trades in the Western states have been from agriculture to municipal and industrial uses (Saliba 1987).

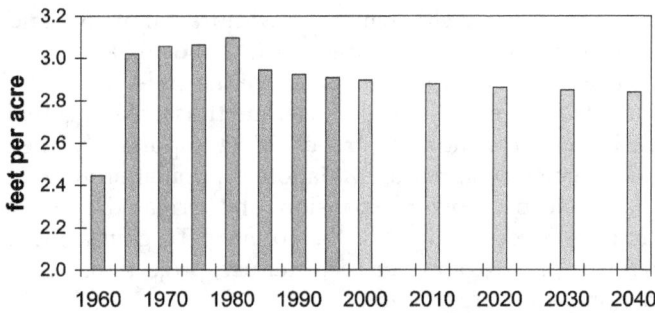

Figure 21. Projected withdrawal per irrigated acre in the West.

to decrease from 514 gallons per day in 1995 to 354 in 2040 assuming the middle series population projections.

Net change in total withdrawal

The graph at the bottom of figure 16.1 shows the net change in projected total withdrawal in comparison with the 1995 level, assuming the middle series population projections. Projected total withdrawal increases by 24 bgd (7%) from 1995 to 2040. The largest increases are in the domestic and public (13 bgd) and thermoelectric (11 bgd) sectors. The livestock and industrial and commercial sectors each contribute another 2 bgd, and irrigation decreases by 4 bgd. On a per-capita basis, total withdrawals are projected to decrease from 1301 gallons per day in 1995 to 992 in 2040.

Holding the overall increase below 10% of total 1995 withdrawals, in spite of the 41% increase in population, is largely attributable to 1) the improving efficiencies projected for the municipal and industrial and thermoelectric sectors, and 2) the reductions in total irrigation withdrawal.

Table 2 compares results based on the middle population series with results based on the low and high series projections. In contrast to the 7% increase in total withdrawals from 1995 to 2040 with the middle series, the low

Table 2. National withdrawal projections for alternative population series, expressed as change from 1995 to 2040 (percent change in parentheses).

	Low series	Middle series	High series
Population (millions)	24 (9%)	107 (41%)	195 (74%)
Withdrawal (bgd)[a]			
Livestock	1 (9%)	2 (41%)	4 (75%)
Domestic & public	3 (10%)	13 (42%)	24 (76%)
Ind. & commercial	−6 (−17%)	2 (6%)	12 (32%)
Thermoelectric	−22 (−17%)	11 (9%)	48 (36%)
Irrigation	−4 (−3%)	−4 (−3%)	−4 (−3%)
Total	−29 (−8%)	24 (7%)	83 (24%)

[a] bgd - billion gallons per day

USDA Forest Service Gen. Tech. Rep. RMRS–GTR–39. 1999

15

and high series yield changes in withdrawal of –8% and 24%, respectively (table 2, figures 16.2 and 16.3).

These national projections ignore the site-specific nature of water availability and use, and thus mask regional variations in withdrawal trends. Most importantly, they fail to depict the significant variations in population growth, thermoelectric power expansion, and irrigated acreage changes across regions of the country. To gain a more realistic picture of projected water use, projections at the regional scale are examined.

Projections for Water Resource Regions

Estimates of population and per-capita income were aggregated from the county level to the WRR level. Table 3 lists the 1995 population and per-capita income estimates, and the percentage changes in these variables projected for 1995 to 2040. Population and per-capita income are projected to increase in all regions. Population increases range from 26% to 75% and per-capita income increases range from 32% to 47%. The largest increases in population are expected in the South and West.

As described above, rates of change in efficiency factors were computed at different geographical scales. For livestock, domestic and public, industrial and commercial, and thermoelectric withdrawals, national-level rates were used; for irrigation withdrawal per acre, East/West-level rates were used; and for irrigated acreage, WRR-level rates were applied. These rates of change in factors affecting withdrawals were applied at the WRR level to beginning levels set equal in most cases to the 1995 level (note exceptions in footnote 10).

For irrigated acreage, the WRR-specific rates of change were chosen to extend recent trends and are always assumed to gradually decrease over time. The effects of these rates of change on WRR irrigated acreage are listed in table 4 and depicted in the plots of irrigated acres in each of the 20 parts of figure 22. The mean 1985 through 1995 application rates are also listed in table 4. As described above, application rates were projected to remain constant in the East and to drop in the West by 2.9% by the year 2040.

Figure 22 presents past and projected withdrawal levels and related water-use determinant levels for the five water-use categories. Figure 22 reveals numerous anoma-

Table 3. Population and per-capita income for water resource regions.

	Population		Annual per-capita income (1990 dollars)	
	1995 (millions)	Percent change 1995-2040	1995 ($1000)	Percent change 1995-2040
Water resource region				
1. New England	13.4	38	22.9	34
2. Mid Atlantic	42.4	29	23.0	37
3. South Atlantic-Gulf	37.6	55	17.6	42
4. Great Lakes	23.5	26	19.9	38
5. Ohio	21.1	30	17.0	42
6. Tennessee	4.3	42	15.6	42
7. Upper Mississippi	22.8	33	18.9	41
8. Lower Mississippi	7.3	33	15.0	45
9. Souris-Red-Rainy	0.7	26	15.8	42
10. Missouri Basin	10.3	40	18.3	43
11. Arkansas-White-Red	8.7	37	16.4	47
12. Texas-Gulf	15.9	44	18.2	44
13. Rio Grande	3.1	47	12.5	45
14. Upper Colorado	0.7	57	15.5	44
15. Lower Colorado	5.3	70	17.2	41
16. Great Basin	2.4	75	15.7	47
17. Pacific Northwest	9.9	53	18.2	40
18. California	32.3	52	21.3	39
19. Alaska	0.6	54	20.9	35
20. Hawaii	1.2	53	20.5	32
United States	261.2	41	19.3	39

16

USDA Forest Service Gen. Tech. Rep. RMRS–GTR–39. 1999

New England (WRR 1)

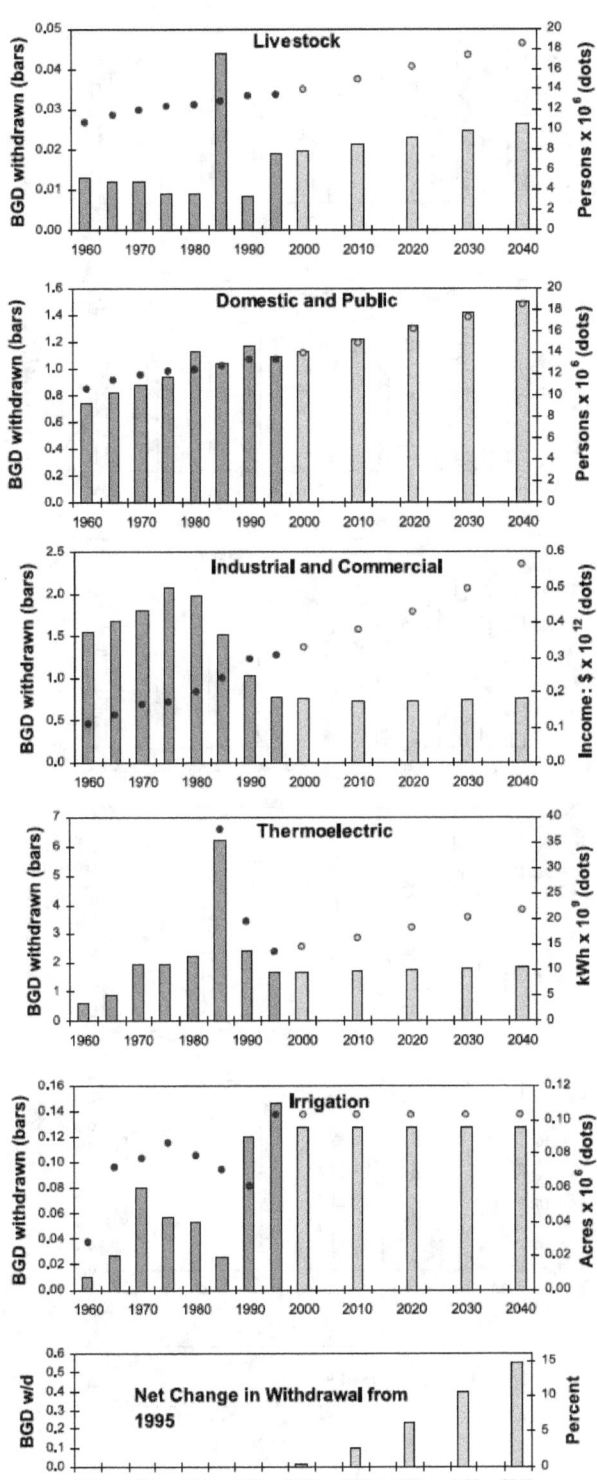

Figure 22.1. Withdrawal projections for New England (Water Resource Region 1).

Mid Atlantic (WRR 2)

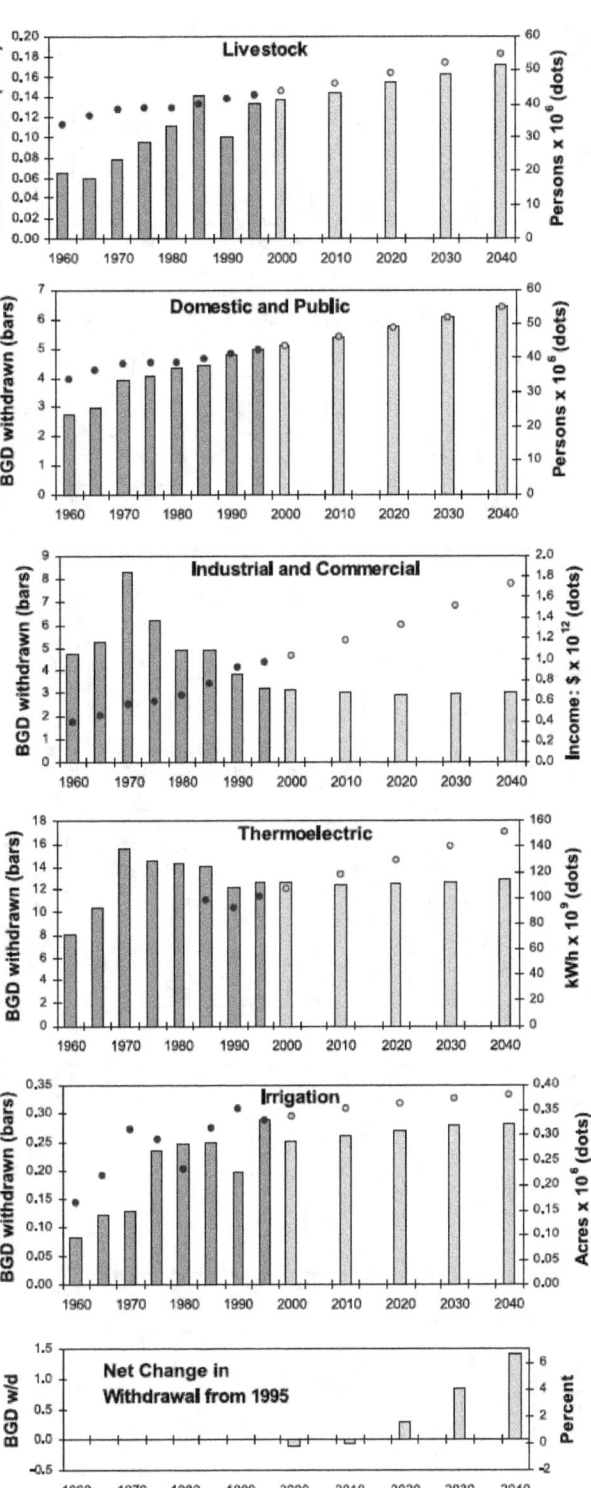

Figure 22.2. Withdrawal projections for Mid Atlantic (Water Resource Region 2).

USDA Forest Service Gen. Tech. Rep. RMRS–GTR–39. 1999

17

South Atlantic-Gulf (WRR 3)

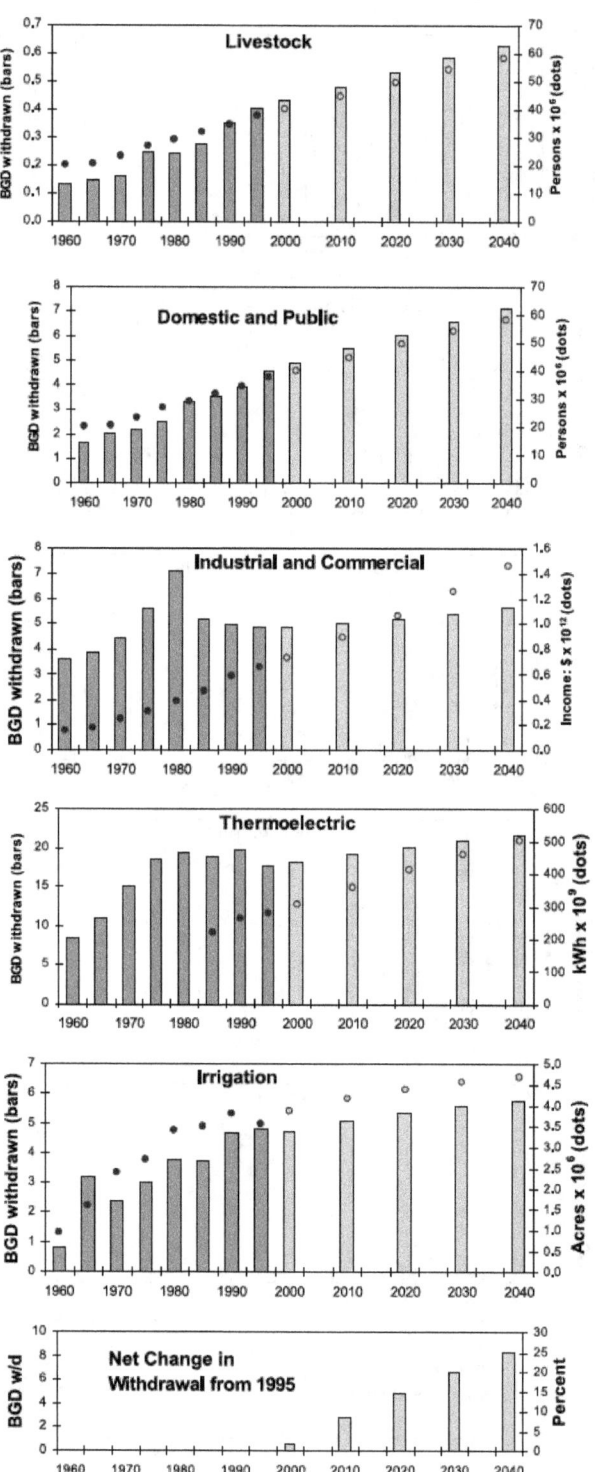

Figure 22.3. Withdrawal projections for South-Atlantic Gulf (Water Resource Region 3).

Great Lakes (WRR 4)

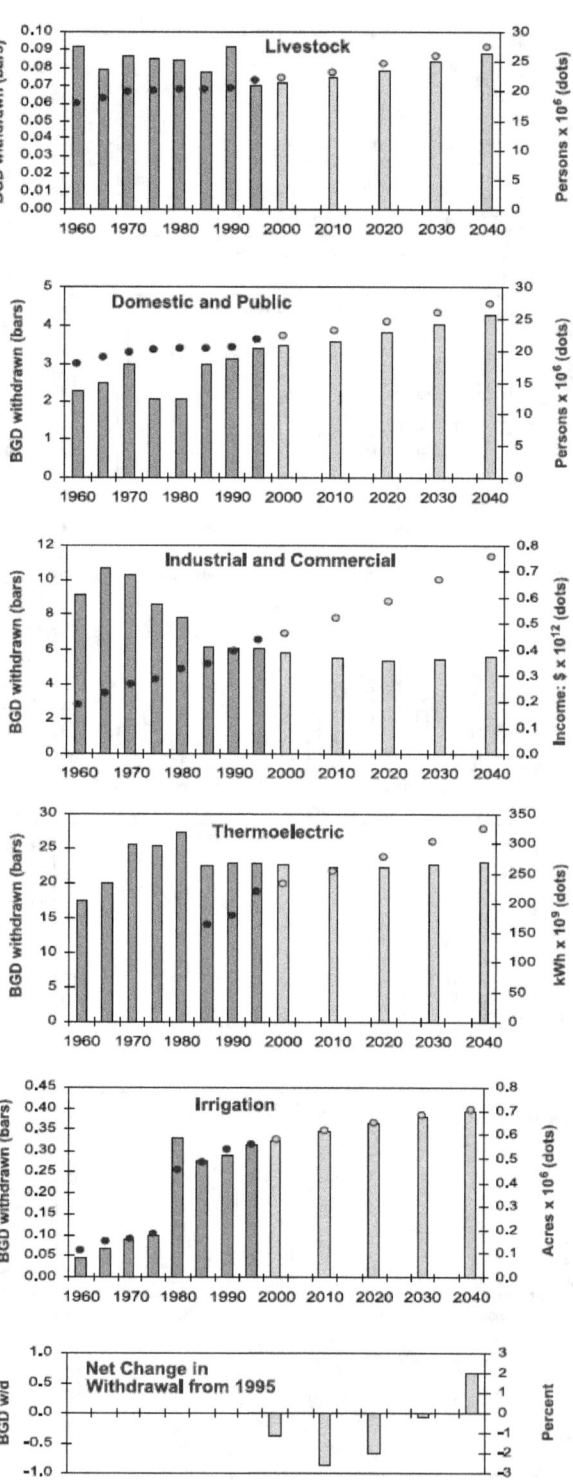

Figure 22.4. Withdrawal projections for Great Lakes (Water Resource Region 4).

18

USDA Forest Service Gen. Tech. Rep. RMRS–GTR–39. 1999

Ohio (WRR 5)

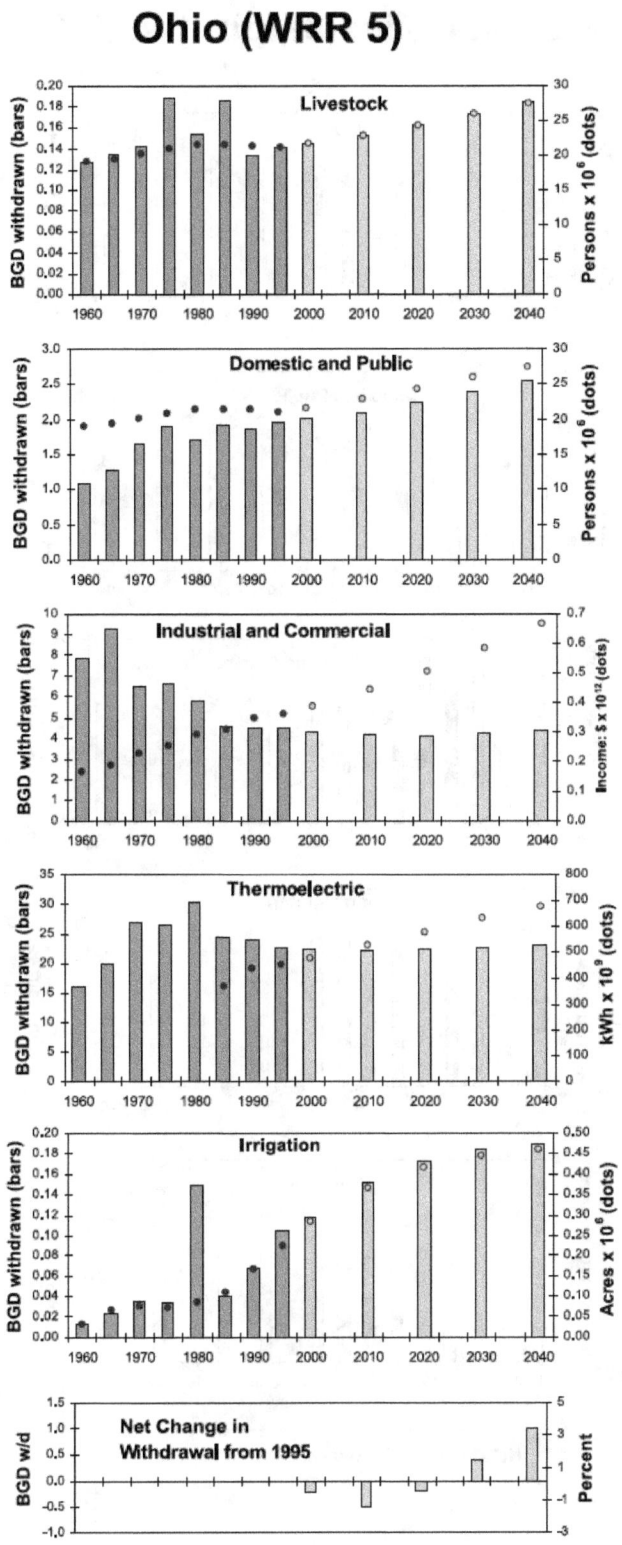

Figure 22.5. Withdrawal projections for Ohio (Water Resource Region 5).

Tennessee (WRR 6)

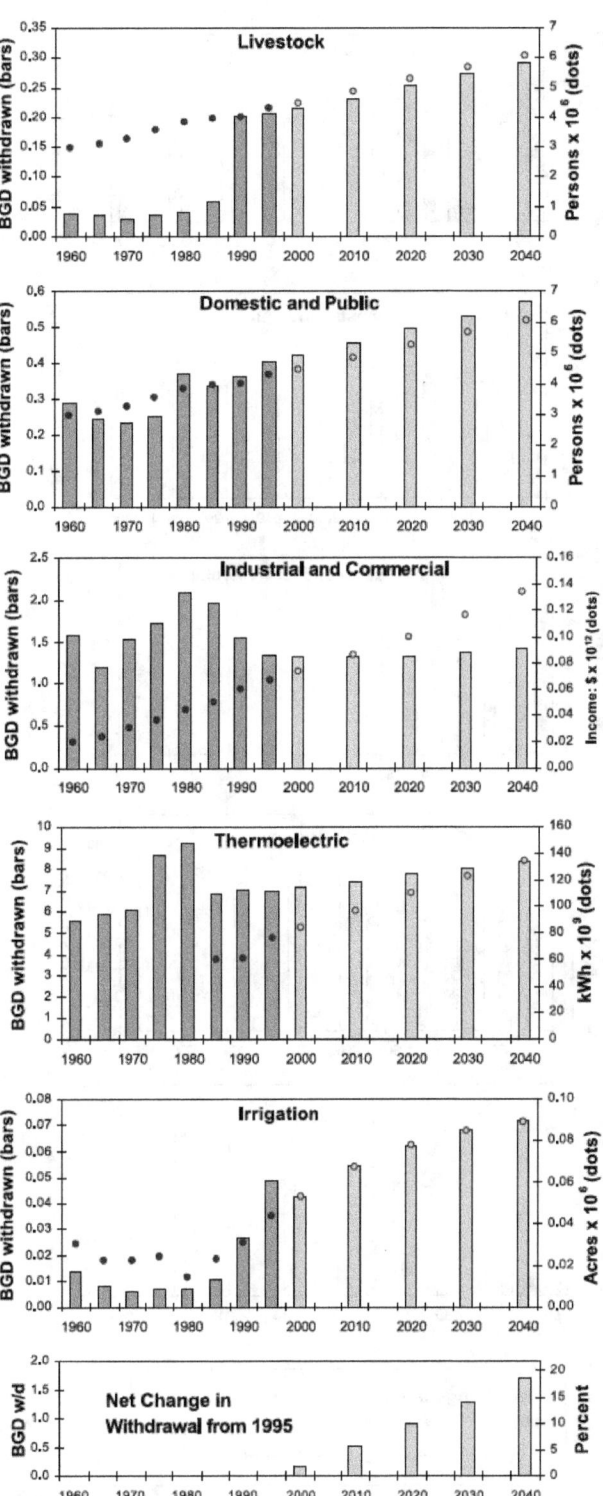

Figure 22.6. Withdrawal projections for Tennessee (Water Resource Region 6).

USDA Forest Service Gen. Tech. Rep. RMRS–GTR–39. 1999

19

Upper Mississippi (WRR 7)

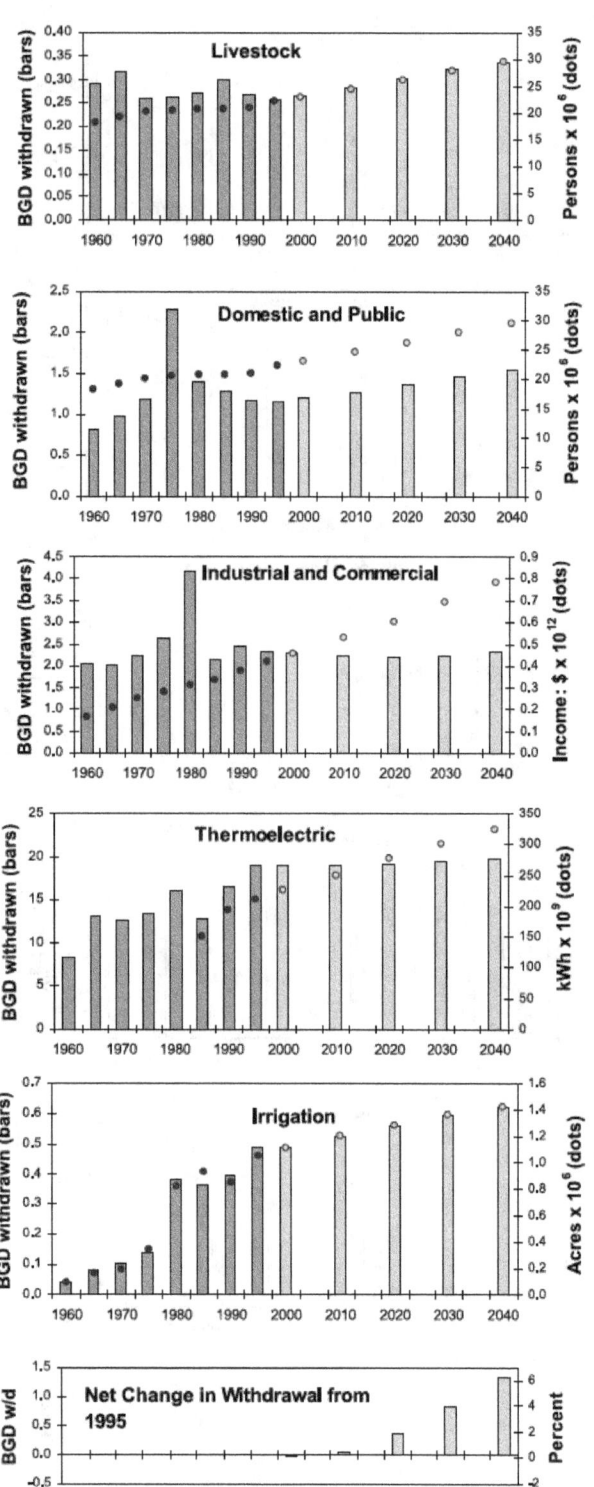

Figure 22.7. Withdrawal projections for Upper Mississippi (Water Resource Region 7).

Lower Mississippi (WRR 8)

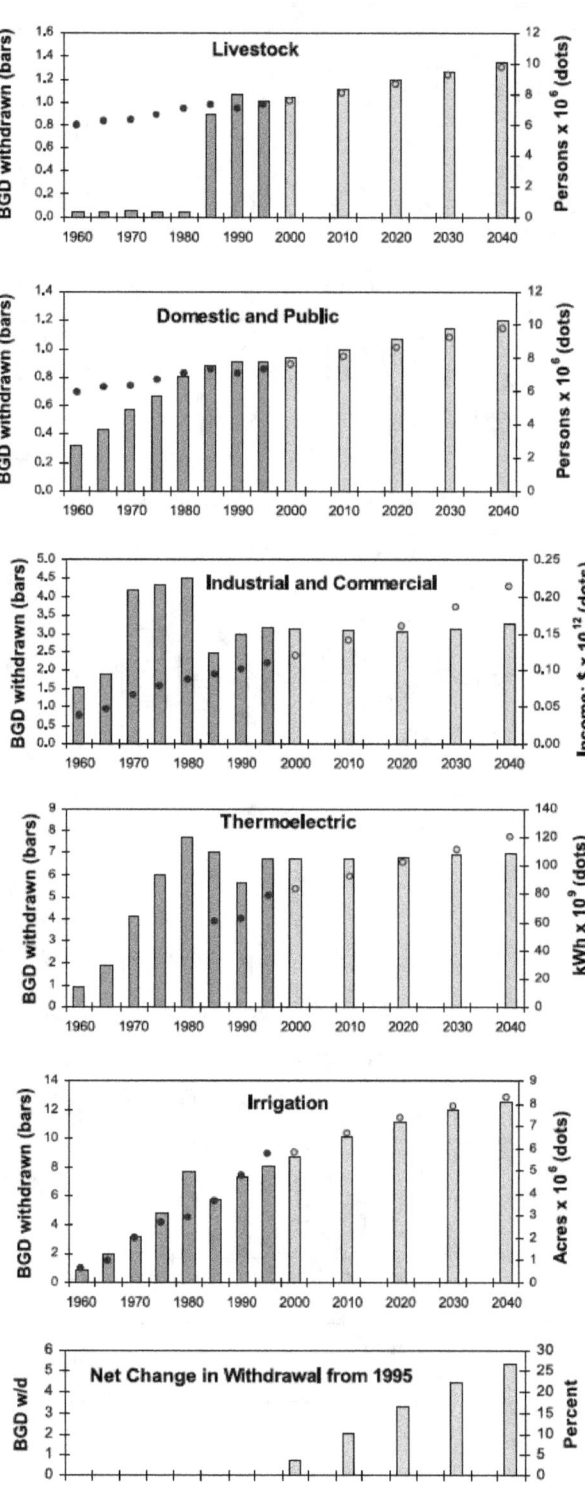

Figure 22.8. Withdrawal projections for Lower Mississippi (Water Resource Region 8).

20

USDA Forest Service Gen. Tech. Rep. RMRS–GTR–39. 1999

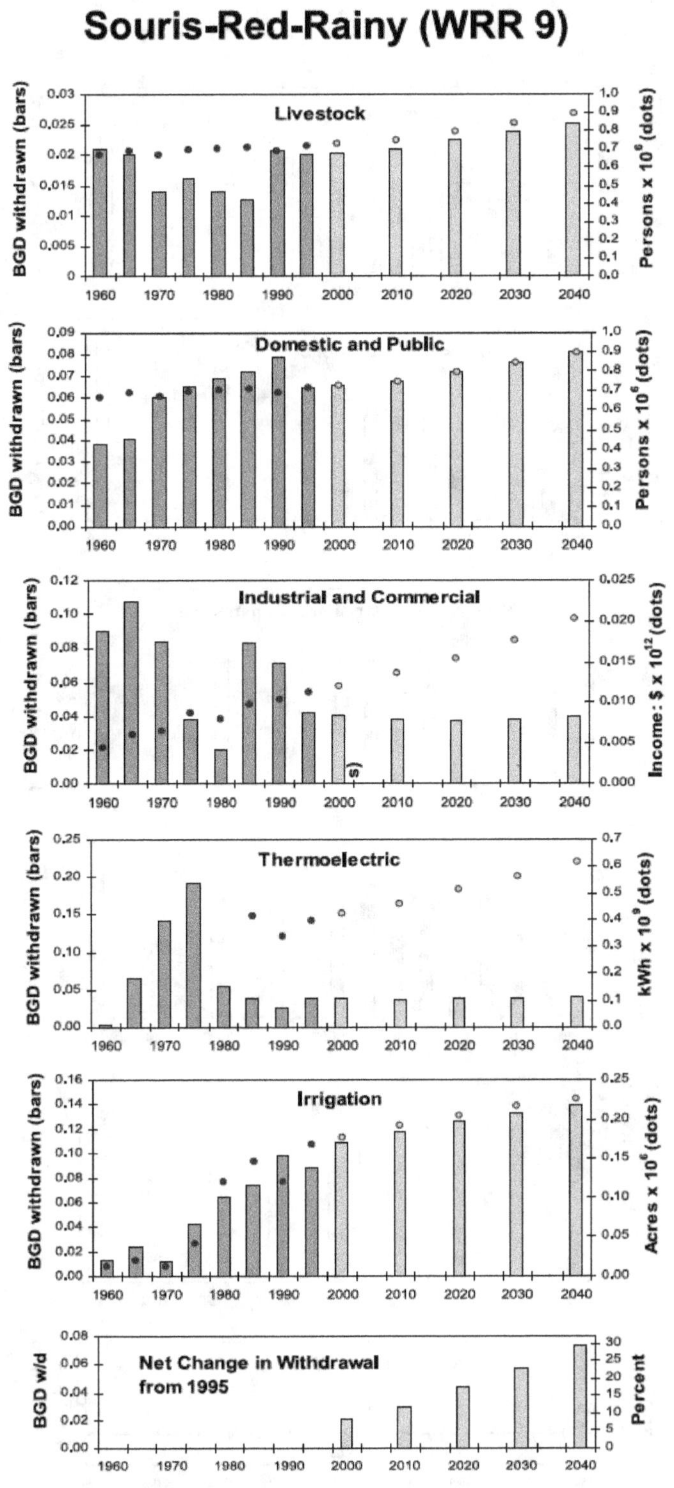

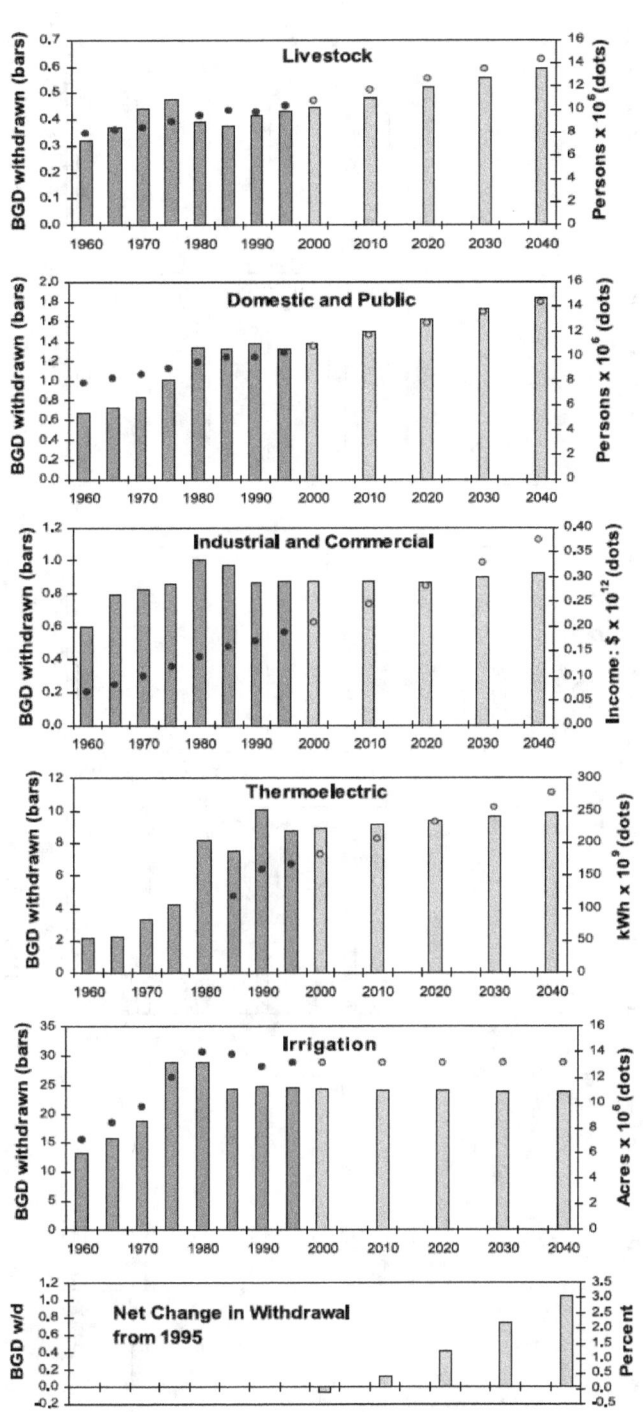

Figure 22.9. Withdrawal projections for Souris-Red-Rainy (Water Resource Region 9).

Figure 22.10. Withdrawal projections for Missouri Basin (Water Resource Region 10).

Arkansas-White-Red (WRR 11)

Texas Gulf (WRR 12)

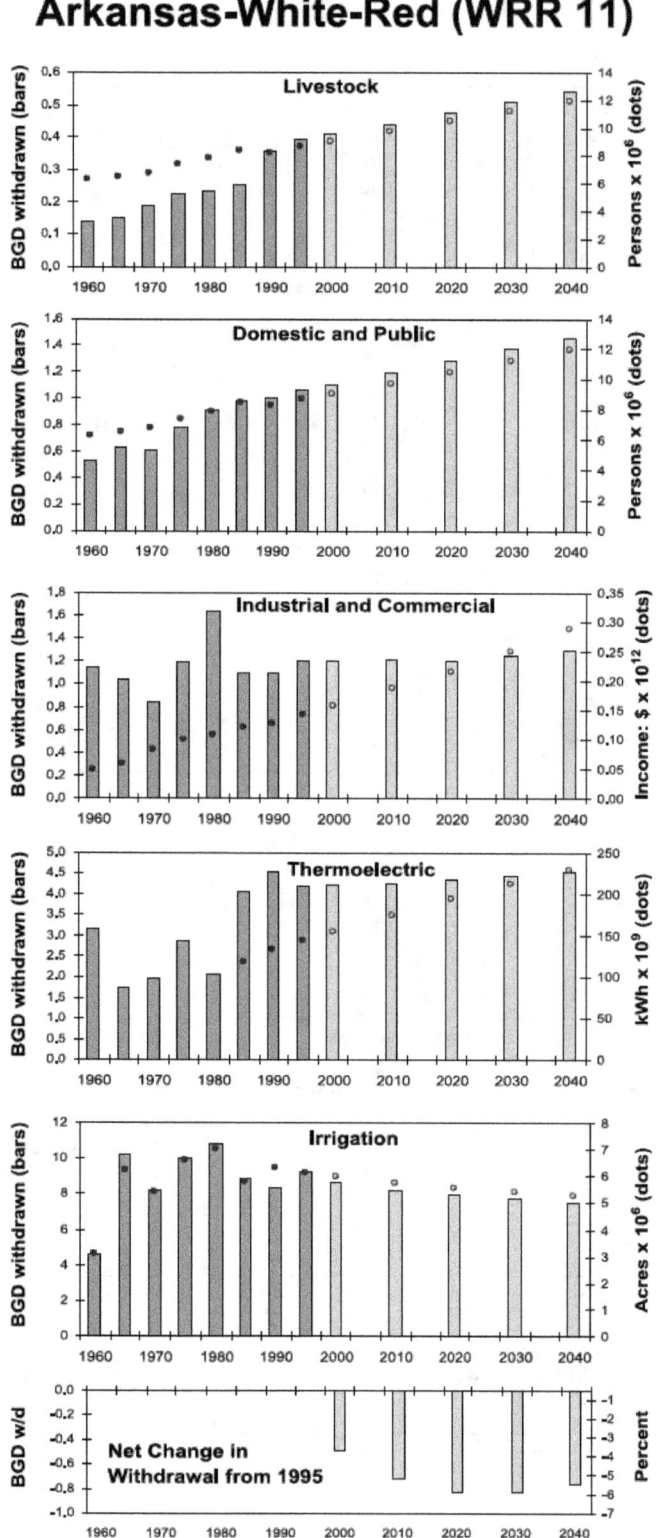

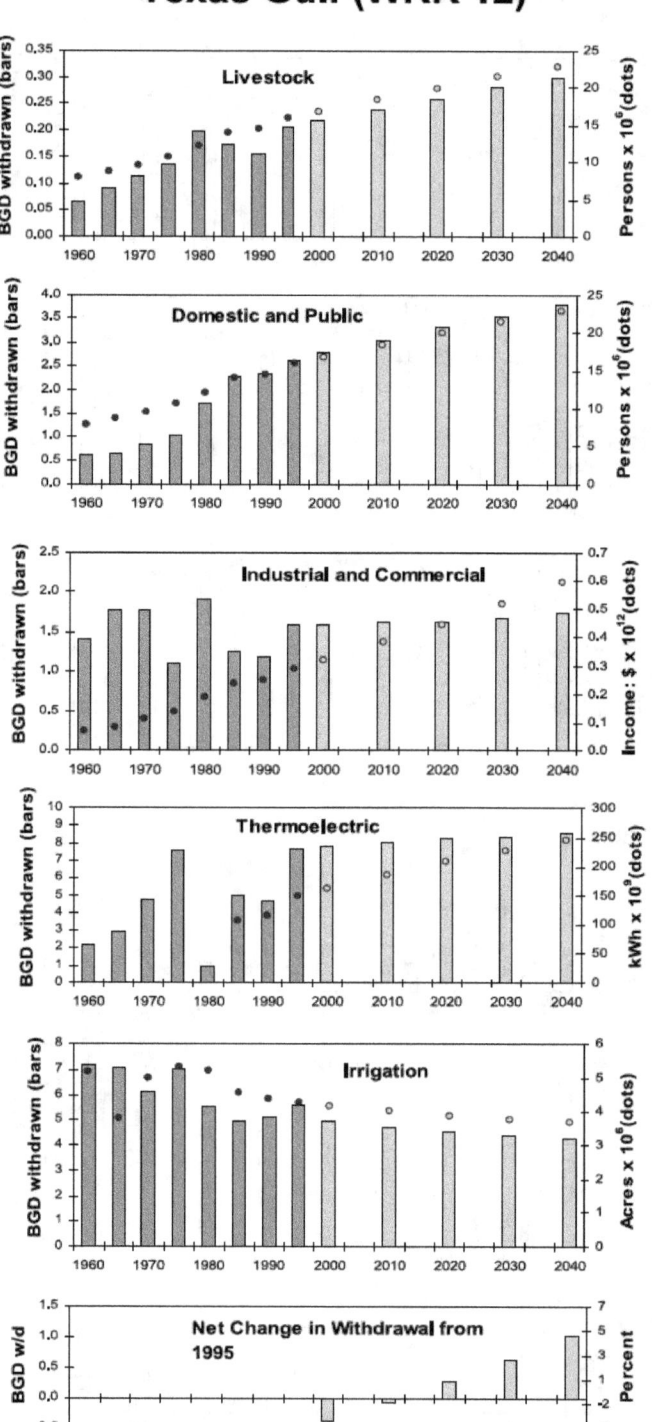

Figure 22.12. Withdrawal projections for Texas-Gulf (Water Resource Region 12).

Figure 22.11. Withdrawal projections for Arkansas-White-Red Water Resource Region 11).

22

USDA Forest Service Gen. Tech. Rep. RMRS–GTR–39. 1999

Rio Grande (WRR 13)

Upper Colorado (WRR 14)

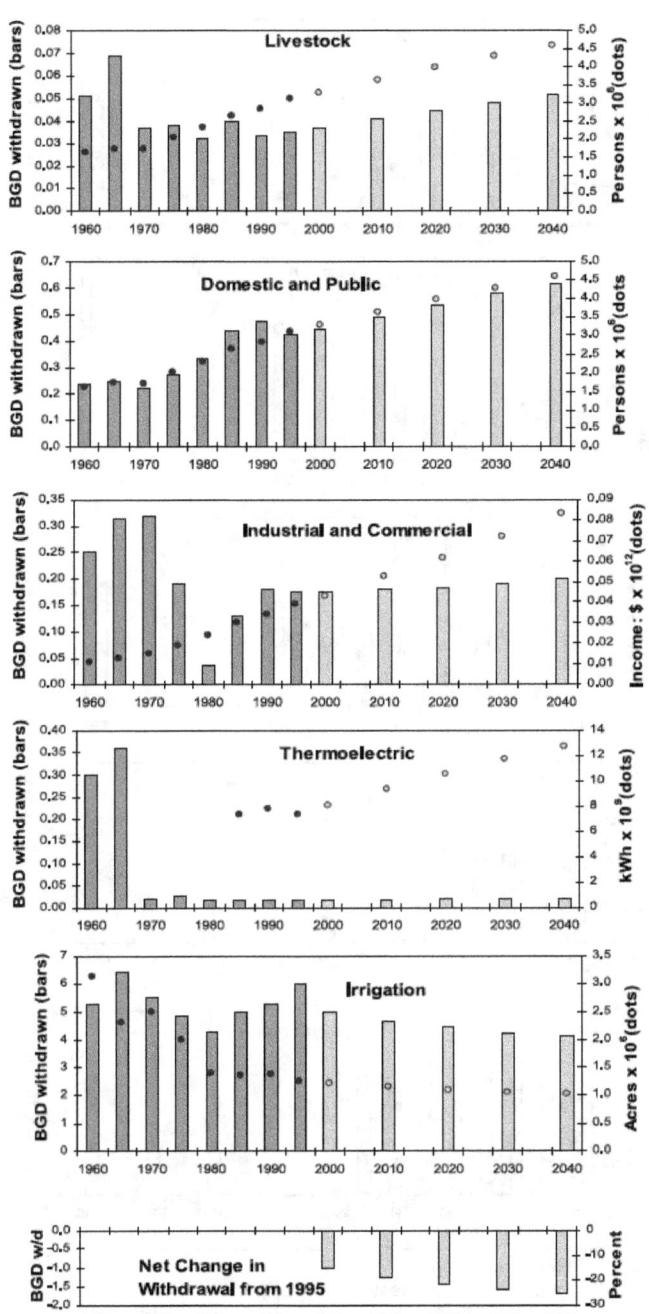

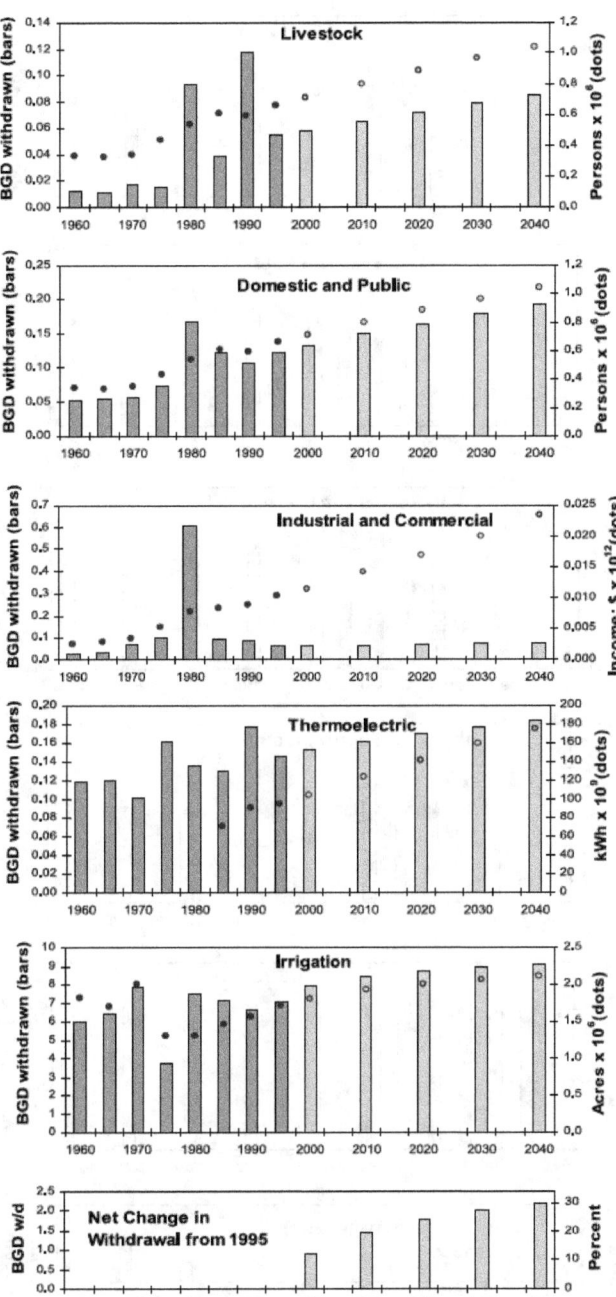

Figure 22.13. Withdrawal projections for Rio Grande (Water Resource Region 13).

Figure 22.14. Withdrawal projections for Upper Colorado (Water Resource Region 14).

USDA Forest Service Gen. Tech. Rep. RMRS–GTR–39. 1999

23

Lower Colorado (WRR 15)

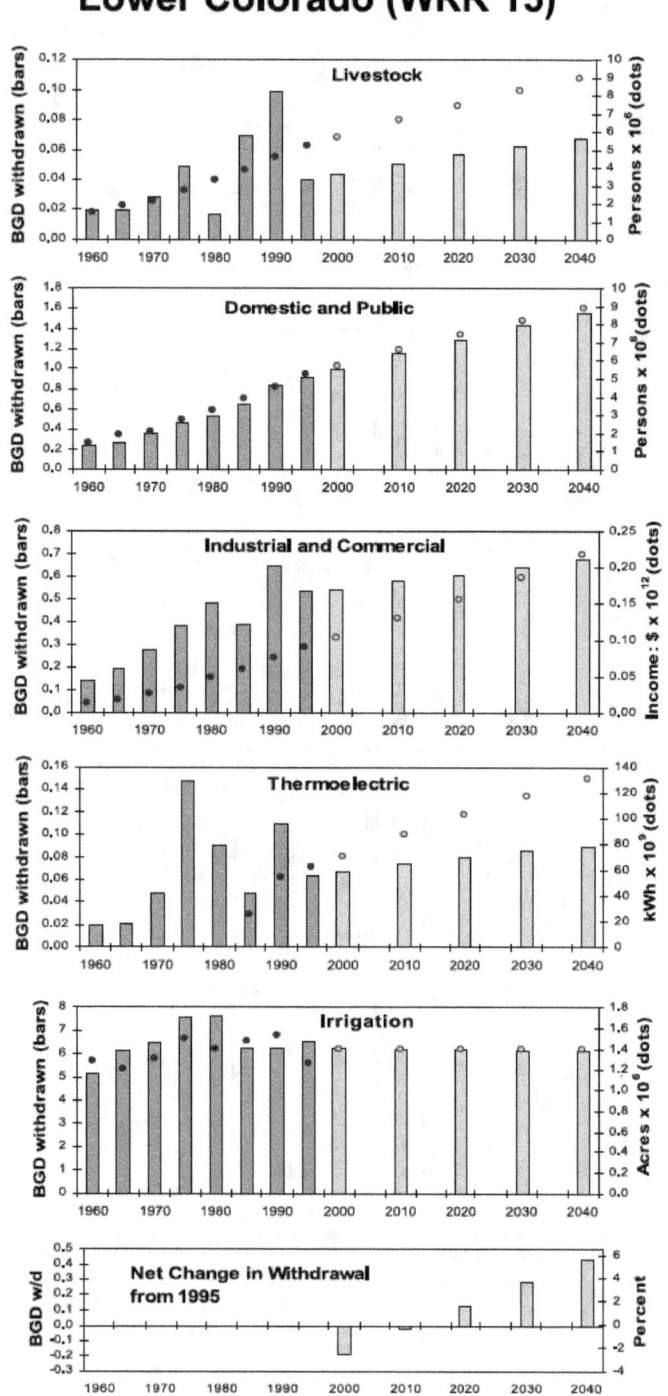

Figure 22.15. Withdrawal projections for Lower Colorado (Water Resource Region 15).

Great Basin (WRR 16)

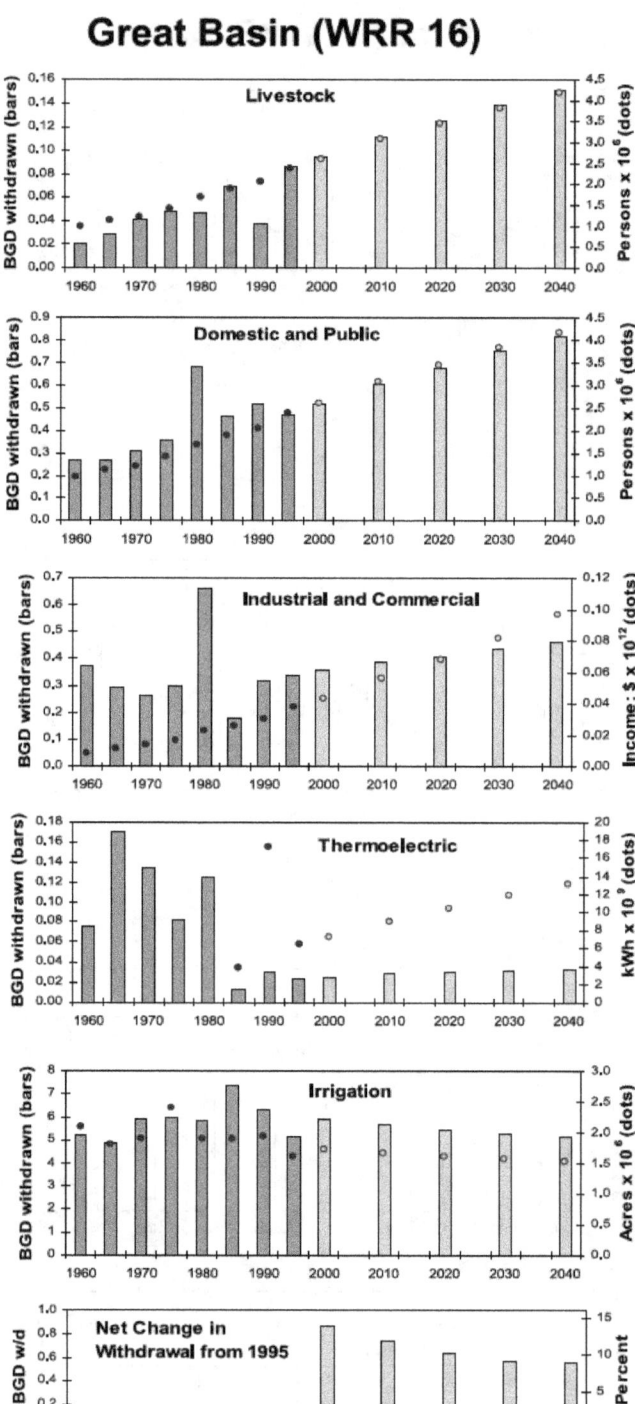

Figure 22.16. Withdrawal projections for Great Basin (Water Resource Region 16).

Pacific Northwest (WRR 17)

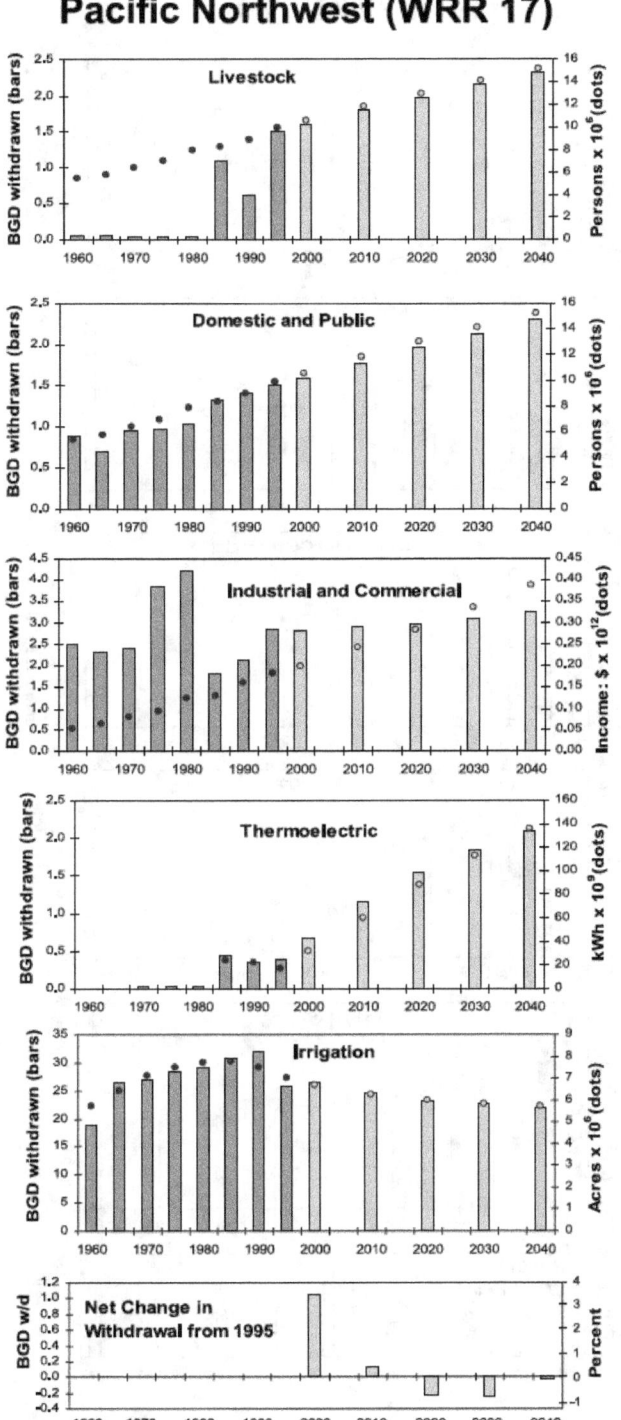

Figure 22.17. Withdrawal projections for Pacific Northwest (Water Resource Region 17).

California (WRR 18)

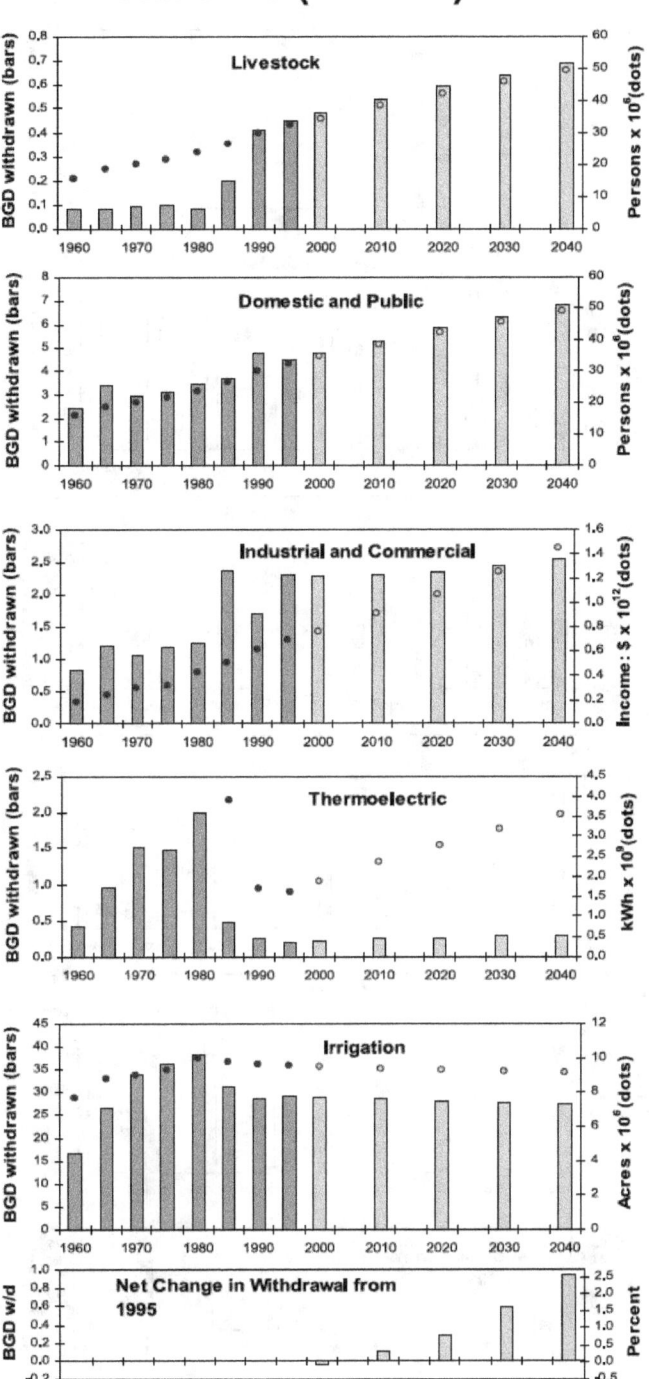

Figure 22.18. Withdrawal projections for California (Water Resource Region 18).

Alaska (WRR 19)

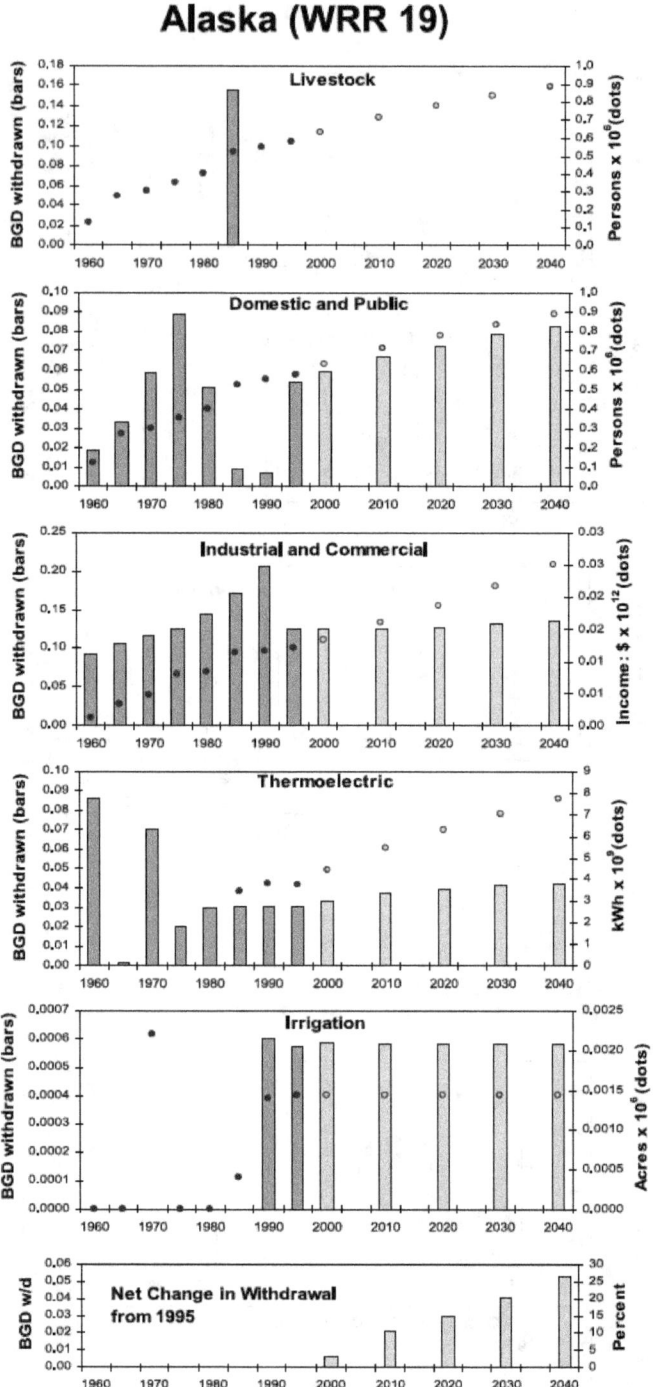

Figure 22.19. Withdrawal projections for Alaska (Water Resource Region 19).

Hawaii (WRR 20)

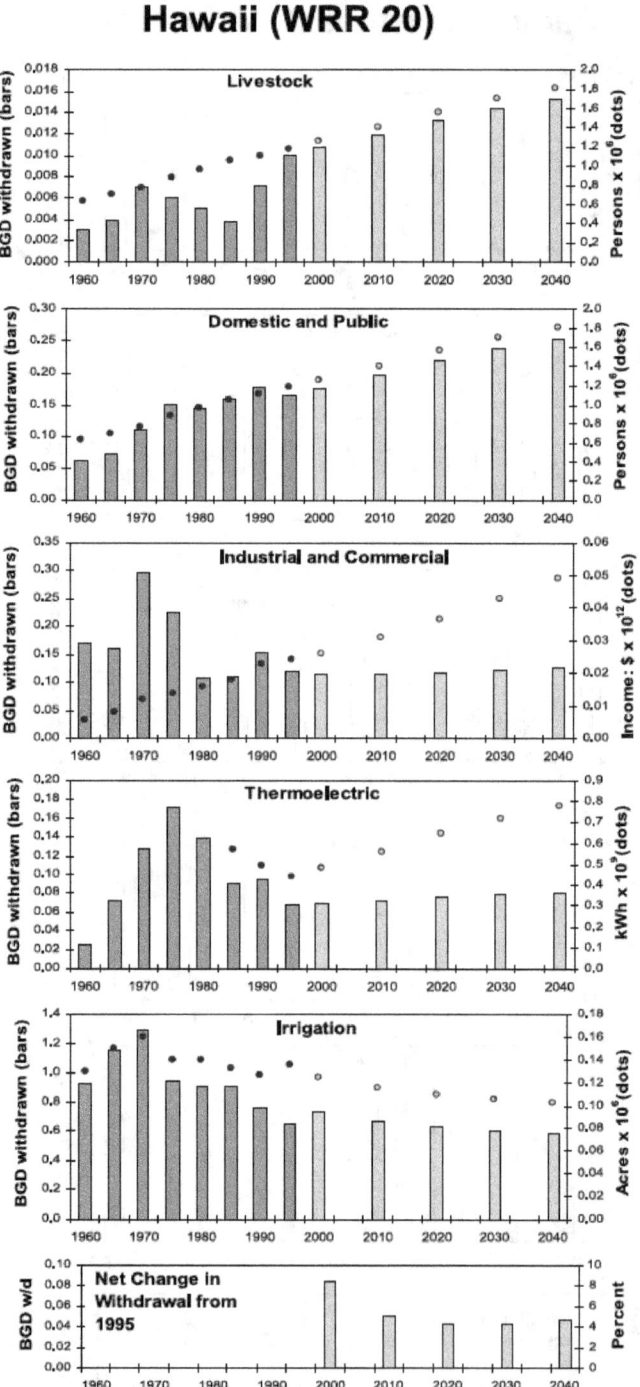

Figure 22.20. Withdrawal projections for Hawaii (Water Resource Region 20).

lies that were not apparent in the aggregated data of figure 16. For example:

- in WRR 1, livestock withdrawal in 1985 is unusually high;

- in WRR 5, irrigation withdrawal in 1980 is unusually high;

- in WRR 7, domestic and public withdrawal in 1970, and industrial and commercial withdrawal in 1980, are unusually high;

- in WRR 8, industrial and commercial withdrawals in 1970, 1975, and 1980 are unusually high;

- in WRR 9, industrial and commercial withdrawals in 1975 and 1980 are unusually low;

- in WRR 13, thermoelectric withdrawals in 1960 and 1965 are unusually high, and industrial and commercial withdrawals in 1980 are unusually low;

- in WRR 14, livestock, domestic and public, and industrial and commercial withdrawals in 1980 are unusually high, and irrigated acreage in 1970 is unusually low;

- in WRR 18, thermoelectric withdrawals in 1985 drop precipitously; and

- in WRR 19, livestock withdrawals in 1985 are unusually high.

Explanations for these unusual entries are often difficult to obtain, especially for earlier years. Some of the entries may be errors, but most probably have logical explanations (for example, in WRR 18, the decrease in thermoelectric withdrawal in 1985 is related to a switch from fresh to salt water). The purpose here is not to explain every entry, but rather to look at overall trends. Nevertheless, the anomalies highlight the importance of not placing too much importance on individual data entries.

Water withdrawal projections by WRR are shown in figure 22 and summarized in tables 5 and 6 and in tables A1.1 to A1.6. All WRRs show increases in withdrawals for livestock and domestic and public uses, in keeping with the assumptions of increasing population in all regions (table 5). On a percentage basis (table 6), the 1995 to 2040 increases in livestock and domestic and public withdrawals are most pronounced in the Western regions and in the South Atlantic Gulf region, where the greatest population increases are expected. Six regions show decreases in

Table 4. Assumptions about future irrigation in the water resource regions.

Water resource region	1985–1995 mean withdrawal (feet per acre)[a]	Irrigated acres	
		1995 (in thousands)	Percent change 1995 to 2040
1. New England	1.40	103	0
2. Mid Atlantic	0.84	328	16
3. South Atlantic-Gulf	1.37	3,552	32
4. Great Lakes	0.63	556	27
5. Ohio	0.46	222	107
6. Tennessee	0.90	44	104
7. Upper Mississippi	0.49	1,054	35
8. Lower Mississippi	1.70	5,730	44
9. Souris-Red-Rainy	0.69	168	35
10. Missouri Basin	2.08	13,163	0
11. Arkansas-White-Red	1.64	6,117	−14
12. Texas-Gulf	1.33	4,279	−14
13. Rio Grande	4.59	1,264	−19
14. Upper Colorado	4.96	1,709	23
15. Lower Colorado	5.06	1,257	11
16. Great Basin	3.87	1,607	−5
17. Pacific Northwest	4.44	7,030	−19
18. California	3.44	9,539	−4
19. Alaska	0.46	1	0
20. Hawaii	6.60	136	−24
United States	2.64	57,857	3

[a] For Alaska, the mean was computed for 1990 and 1995.

Table 5. Change in withdrawal from 1995 to 2040, middle population series, in million gallons per day.

Water resource region	Livestock	Domestic & public	Indust. & commercial	Thermo-electric	Irrigation	Total
1. New England	7	412	−18	167	−18	550
2. Mid Atlantic	39	1,456	−208	140	−5	1,422
3. South Atlantic-Gulf	223	2,521	807	3,936	919	8,407
4. Great Lakes	18	866	−481	185	79	668
5. Ohio	43	593	−91	381	85	1,010
6. Tennessee	86	168	87	1320	23	1,684
7. Upper Mississippi	84	385	−17	742	135	1,330
8. Lower Mississippi	333	298	73	262	4,410	5,377
9. Souris-Red-Rainy	5	17	−2	2	51	73
10. Missouri Basin	169	524	51	1,100	−783	1,061
11. Arkansas-White-Red	147	397	85	354	−1,751	−768
12. Texas-Gulf	91	1,157	150	914	−1,290	1,023
13. Rio Grande	17	198	23	3	−1,922	−1,682
14. Upper Colorado	31	70	13	38	2,068	2,219
15. Lower Colorado	28	641	145	27	−407	433
16. Great Basin	65	353	123	9	7	558
17. Pacific Northwest	804	797	389	1,705	−3,718	−24
18. California	235	2,306	262	100	−1,956	947
19. Alaska	0	29	12	12	0	53
20. Hawaii	5	88	9	14	−69	46
United States	2,430	13,275	1,411	11,411	−4,142	24,386

Table 6. Percent change in withdrawal from 1995 to 2040, middle population series.

Water resource region	Livestock	Domestic & public	Indust. & commercial	Thermo-electric	Irrigation	Total
1. New England	38	38	−2	10	−12	15
2. Mid Atlantic	29	29	−6	1	−2	7
3. South Atlantic-Gulf	55	55	17	22	19	26
4. Great Lakes	26	26	−8	1	25	2
5. Ohio	30	30	−2	2	81	3
6. Tennessee	42	42	6	19	48	19
7. Upper Mississippi	33	33	−1	4	28	6
8. Lower Mississippi	33	33	2	4	54	27
9. Souris-Red-Rainy	26	26	−5	5	58	29
10. Missouri Basin	40	40	6	12	−3	3
11. Arkansas-White-Red	37	37	7	8	−19	−5
12. Texas-Gulf	44	44	9	12	−23	6
13. Rio Grande	47	47	13	17	−32	−25
14. Upper Colorado	57	57	20	26	29	30
15. Lower Colorado	70	70	27	42	−6	5
16. Great Basin	75	75	36	39	0	9
17. Pacific Northwest	53	53	14	443	−14	0
18. California	52	52	11	49	−7	3
19. Alaska	54	54	10	39	1	25
20. Hawaii	53	53	7	20	−11	5
United States	41	42	6	11	−3	7

withdrawals for industrial and commercial uses (ranging from 1% to 8%), despite the expected increases in economic activity, because of the assumed increase in efficiency of industrial and commercial water use. In the other 14 regions, population and per-capita income increases overwhelm the increasing efficiencies of water use to cause projected increases in industrial and commercial withdrawals of from 2% to 36% (table 6).

Withdrawals for thermoelectric plants are greater in 2040 than in 1995 in all regions (table 5). However, withdrawals in about half of the regions peak in 2020 or 2030 (figure 22). The largest percentage increases occur in regions of greatest population increase and where production at freshwater plants is currently relatively less important (where a large portion of current production occurs at hydroelectric plants). The very large percentage increase in the Pacific Northwest is attributable to the fact that hydroelectric dams now provide 89% of the region's electricity production, a far greater percentage than for any other region. With the assumptions of constant hydroelectric production at the 1995 level and constant proportional allocation of thermoelectric production between freshwater and saltwater plants, production at freshwater thermoelectric plants must increase substantially to accommodate the required increase in total electric energy production in the Pacific Northwest of 87% from 1995 to 2040.

Irrigation withdrawals are projected to increase in 9 regions, decrease in 10, and remain constant in 1 (table 5). Only three regions experience substantial volume increases in withdrawal (the South Atlantic-Gulf, Lower Mississippi, Upper Colorado regions), although in percentage terms several other regions also show substantial increases (table 6). The changes are largely a function of the assumed irrigated acreage changes, although the changes in application per acre in the West also play a role.

Changes in total withdrawal are shown in the bottom graphs of figures 22.1 to 22.20 and summarized in the total column of tables 5 and 6. Only 2 regions (Arkansas-White-Red and Rio Grande) are projected to experience decreases in total withdrawal from 1995 to 2040. In these regions, large decreases in irrigation withdrawal outweigh increases in the other four water-use categories. Increases in total withdrawal among the other 18 regions range up to 30% (table 6). Total withdrawal increases exceed 15% in seven regions, which fall into 2 groups based on the water-use categories to which the withdrawal increases would be delivered. In five regions (New England, South Atlantic-Gulf, Tennessee, Great Basin, and Alaska), the increases are largely attributable to increases in domestic and public or thermoelectric uses or both (table 5). In the other three regions (Lower Mississippi, Souris-Red-Rainy, and Upper Colorado), the total withdrawal increases are largely due to irrigation.

Effects of withdrawal increases are felt within a region and perhaps, depending on its location, downstream of the region. Of the eight regions of greatest percentage increase in total withdrawal, outflows from four (New England, South Atlantic-Gulf, Lower Mississippi, and Alaska) flow into the sea. Outflows from the Tennessee Region flow into the Mississippi River and thus are not of overwhelming concern. Outflows from the Souris-Red-Rainy region flow north to Canada. The Great Basin region is closed. But outflows from the Upper Colorado Basin (WRR 14), the region of the largest percentage increase (30%), are of critical concern to 2 downstream basins.

The Colorado River Compact of 1922 limits withdrawals in the Upper Colorado Basin to those that will allow delivery of 8.23 million acre-feet (maf) per year on average to the Lower Basin states of Arizona, Nevada, and California (MacDonnell and others 1995). Tree ring studies indicate that the long-term annual virgin flow at the Lees Ferry delivery point is about 13.5 maf (Stockton and Jacoby 1976).[16] Given this virgin flow and assuming current annual Upper Basin consumptive use of about 3.5 maf, the resulting average annual deliveries to the Lower Basin are about 10 maf, far above the Compact requirement of 8.23 maf.[17] With the projected increase in Upper Basin withdrawal of 2.1 bgd (2.7 maf) (table 5), and assuming that 34% of withdrawals are consumptively used (table A1.9), Upper Basin depletions would increase to 4.3 maf by 2040. These depletions would leave average annual deliveries to the Lower Basin of 9.2 maf, which is still roughly 1 maf above the Compact requirement.

Because much of the current surplus in delivery to the Lower Basin is typically diverted to Southern California, reducing the average annual delivery to the Lower Basin to 9.1 maf would cause some reductions in actual Lower Basin withdrawals. This situation would be exacerbated as Lower Basin demands also increase with population growth (see figures 22.15 and 22.18). Further, these figures apply to the average year; in times of successive dry years, when Colorado River storage is drawn down, Upper Basin withdrawals may be curtailed in order to meet the Compact delivery requirement.

---

[16] *Lees Ferry is located near the boundary between the Upper and Lower Colorado Basins, just upstream of the Grand Canyon. Annual virgin flows at Lees Ferry during this century have averaged 16.5 maf, but this average was amplified by unusually high flows during the century's first three decades.*

[17] *Solley and others (1998) estimate 1995 Upper Colorado Basin consumptive use at 2.82 maf. Adding 0.65 maf of reservoir evaporation (Brown and others 1990) brings total consumptive use to 3.5 maf. This estimate is lower than some others. For example, based on U.S. Bureau of Reclamation estimates, Harding and others (1995) estimated average year total depletions to be 4.3 maf, and Brown and others (1990) estimated them to be 4.4 maf.*

Of the seven remaining WRRs of greatest percentage increase in total withdrawal, Alaska is of least concern because water supplies exceed the relatively low levels of withdrawal in that region. However, the increase in total withdrawal is of concern in the other six regions because of within-region impacts. Although conditions in most of the regions are quite humid, water supplies can be strained during relatively dry times. And the impacts will not be felt uniformly across a region. The percentage estimates presented here are statistical averages that are unlikely to apply in any one place. The impacts in individual locations will vary above and below the regional averages depending on local supply and demand characteristics. Further, impacts are not limited to the categories of withdrawal discussed here. Wherever diversions increase, instream flow and surface water quality can be expected to decrease by the consumptive use percentage, all else equal.

Sensitivity of Projections to Assumptions About Factors Affecting Water Use

For the purpose of sensitivity analysis, the projections of water-use factors presented above (factors are listed in table 1), including the middle population series, are called best-guess estimates, and the related water withdrawal projections are best-guess projections. For example, the best-guess withdrawal projection for the U.S. is that total withdrawals will increase from 340 bgd in 1995 to 364 bgd in 2040, a 7% increase. The best-guess projections are in figures 16.1 and 22 and listed in tables 5 and 6 and tables A1.1 to A1.6. This section investigates the effect on withdrawal projections of altering the best-guess estimates of the water-use factors.

Table 7 compares the best-guess withdrawal projections with those using the low and high population series estimates, in terms of the percentage change from 1995 to 2040. Figures 16.2 and 16.3 depict U.S. withdrawals with the low and high population series, for comparison with the best-guess projections in figure 16.1. With the low series, total withdrawals drop by 8% for the U.S. as a whole; efficiencies in industrial and commercial and thermoelectric water uses more than compensate for the effect of the 9% increase in total population from 1995 to 2040. The decrease in total withdrawal occurs in all WRRs except those where significant increases in irrigation withdrawals are projected. Conversely, with the high series, total U.S. withdrawals are projected to increase by 24% from 1995 to 2040. Increases in total withdrawal are projected for all but the Rio Grande region, where the drop in irrigation withdrawal outweighs the increases caused by growing population. In the other 19 regions, total withdrawals would increase by at least 7%, with 12 regions experiencing increases of at least 25%. This high popula-

Table 7. Percent change in total withdrawal from 1995 to 2040 for three Census Bureau population projection series.

Water resource region	Low	Middle	High
1. New England	−11	15	42
2. Mid Atlantic	−17	7	32
3. South Atlantic-Gulf	2	26	52
4. Great Lakes	−22	2	28
5. Ohio	−20	3	28
6. Tennessee	−10	19	50
7. Upper Mississippi	−18	6	31
8. Lower Mississippi	13	27	42
9. Souris-Red-Rainy	12	29	47
10. Missouri Basin	−6	3	12
11. Arkansas-White-Red	−16	−5	7
12. Texas-Gulf	−12	6	25
13. Rio Grande	−28	−25	−22
14. Upper Colorado	28	30	32
15. Lower Colorado	−1	5	12
16. Great Basin	4	9	15
17. Pacific Northwest	−9	0	9
18. California	−4	3	9
19. Alaska	−4	25	56
20. Hawaii	−6	5	16
United States	−8	7	24

tion increase would place extreme pressure on water resources, with serious consequences for offstream users, instream flow, and water quality.

Table 8 lists the percent change, from the best-guess scenario, in total withdrawal in 2040 that is caused by a 10% change in factors affecting projected water use. For example, the lower left estimate of 6.6% indicates that if the future U.S. population were 10% greater than projected by the Census Bureau's middle series, total withdrawals in year 2040 would be 6.6% greater than those indicated in figure 16.1 or tables 5 and A1.6. The effects of changes in the factors are considered separately, such that each column of the table investigates a single change, with all other factors remaining at their best-guess estimates.

For the U.S. as a whole, a 10% increase in population has a larger impact on withdrawals than a 10% change in any of the other six factors listed in table 8. For example, a 10% increase in withdrawal per person causes only a 1.2% increase in total withdrawal, and a 10% increase in kilowatt hours per person causes a 4.1% increase in total withdrawal. The relatively large effect of population change is due partially to the fact that population is a variable in projections of withdrawals of four of the five water-use categories; only future irrigation withdrawals were not modeled as a function of population.

Table 8. Percent change in total withdrawal from best-guess scenario in year 2040 caused by a 10% increase in factors affecting water use.

Water resource region	Population	Factor				
		Withdrawal per person	Withdrawal per $PCI	Withdrawal per kWh	KWh per person	Acres irrigated
1. New England	9.9	3.5	1.8	4.3	4.5	0.3
2. Mid Atlantic	10.0	2.8	1.3	5.6	5.7	0.1
3. South Atlantic-Gulf	8.7	1.7	1.4	5.3	5.4	1.4
4. Great Lakes	10.4	1.3	1.7	6.9	7.4	0.1
5. Ohio	10.0	0.8	1.4	7.6	7.7	0.1
6. Tennessee	10.9	0.5	1.3	7.8	8.7	0.1
7. Upper Mississippi	9.8	0.6	0.9	8.0	8.1	0.3
8. Lower Mississippi	5.1	0.5	1.3	2.8	2.8	5.0
9. Souris-Red-Rainy	5.9	2.5	1.2	1.2	1.4	4.3
10. Missouri Basin	3.7	0.5	0.2	2.6	2.8	6.4
11. Arkansas-White-Red	5.1	0.9	0.8	2.9	3.0	4.9
12. Texas-Gulf	7.7	2.0	0.9	4.6	4.6	2.3
13. Rio Grande	1.8	1.2	0.4	0.0	0.0	8.2
14. Upper Colorado	0.6	0.2	0.1	0.2	0.2	9.4
15. Lower Colorado	2.8	1.8	0.8	0.1	0.1	7.2
16. Great Basin	2.2	1.2	0.7	0.0	0.1	7.8
17. Pacific Northwest	3.7	0.7	1.0	0.6	1.3	6.9
18. California	2.7	1.8	0.7	0.1	0.1	7.3
19. Alaska	10.3	3.2	5.2	1.6	1.9	0.0
20. Hawaii	4.4	2.3	1.2	0.7	0.8	5.5
United States	6.6	1.2	1.1	3.9	4.1	3.6

Comparison across regions indicates that sensitivity to the various changes in assumptions is greatest in regions where the affected water-use accounts for a larger share of total withdrawal. For example, the effect of a percentage increase in domestic and public withdrawal per person is greatest in WRR 1 (New England), where domestic and public uses are relatively more important (accounting for 35% of total withdrawals in 2040). The effect of a percentage increase in irrigated acres is greatest in WRRs 13 (Rio Grande) and 14 (Upper Colorado), where irrigation is more important (accounting for 82% and 94%, respectively, of total 2040 withdrawals).

The sensitivity of total withdrawals to a 10% increase in factors affecting thermoelectric withdrawals is very small in most of the Western regions. In WRRs 13 through 16, the principal reason for the low sensitivity is the small amount of water withdrawn per kilowatt hour produced. Withdrawal in these regions tends to be less than 1 gallon per kWh, whereas in most other regions withdrawal tends to range from 10 to 50 gallons per kWh. In WRRs 18 and 20, the low sensitivity is due mainly to the relatively small amount of electricity that would be produced at freshwater thermoelectric plants (2% and 7%, respectively, of total electricity production). In WRR 17, the low sensitivity is due to the combination of a rather low withdrawal rate (8 gallons per kWh) and moderate proportion (52%) of

total electricity produced at freshwater thermoelectric plants.

Table 8 shows the effect on total withdrawals of changes in the factors affecting water use, but does not indicate which changes are more likely to occur. Of the various factors considered in this report, irrigated acreage is perhaps the most difficult to predict because it depends on many factors, some of which (like international trade) are quite volatile. Increasing worldwide food needs, for example, may increase crop prices and lead to increased production by U.S. farmers, some of which would occur on irrigated fields. Considering this, we take a second look at irrigation withdrawals by examining the effect of removing the future decreases in irrigated acreage assumed for the seven WRRs, all in the West, where acreage decreases were projected (table 4).

If irrigated acreage in those seven Western WRRs were to remain constant at 1995 levels, instead of dropping as assumed, total 1995 to 2040 withdrawal changes would be affected as listed in table 9. For example, if irrigated acreage in WRR 11 (Arkansas-White-Red) were to remain constant at the 1995 level, total year 2040 withdrawals in the region would increase by 3% above the 1995 level instead of decreasing by 5% as projected under the best-guess scenario. Only in WRR 13 (Rio Grande) would total withdrawals in 2040 remain below the 1995 level. For the entire U.S., total withdrawal in 2040 would be 10% above the 1995 level rather than 7% as projected for the best-guess scenario.

Domestic and public withdrawals are also difficult to project because of the recent shift in per-capita withdrawal (figure 18). In a compromise between two opposing but plausible possibilities, it was assumed that future per-capita withdrawals would remain constant at close to the 1995 level. However, it could be argued either that the drop in per-capita withdrawal in 1995 was an aberration and the consistent 1960 to 1990 trend of increasing per-capita withdrawal will reestablish itself, or that the 1995 shift is the beginning of a long-term downward trend in response to new national plumbing fixture standards, conservation efforts of water utilities, and other changes. Figure 23 shows these two alternative possibilities for domestic and public withdrawals per capita, for comparison with the best-guess assumption in figure 18. Table 10 lists the effect of these two possibilities on national domestic and public withdrawals and on total withdrawals. Assuming that per-capita domestic and public withdrawals drop to 115 gallons per day by 2040 (low-level assumption), U.S. domestic and public withdrawals rise to 42 bgd instead of 45 bgd under the best-guess assumption. Alternatively, if per-capita domestic and public withdrawals rise to 134 gallons per day by 2040 (high-level assumption), U.S. domestic and public withdrawals rise to 49 bgd. Total U.S. withdrawals rise from 340 bgd in 1995 to 361, 364, or 368 bgd, respectively, with the low, best-guess, and high projections of per-capita domestic and public withdrawals.

Finally, consider what changes in use rates would be required to keep withdrawals of the respective water uses at their 1995 levels, assuming population increased according to the middle series projections. Table 11 lists these changes for the U.S. for five factors affecting water use, focusing on the year 2040. First, consider per-capita

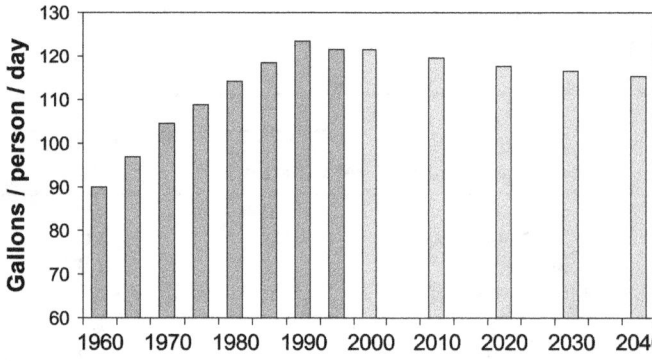

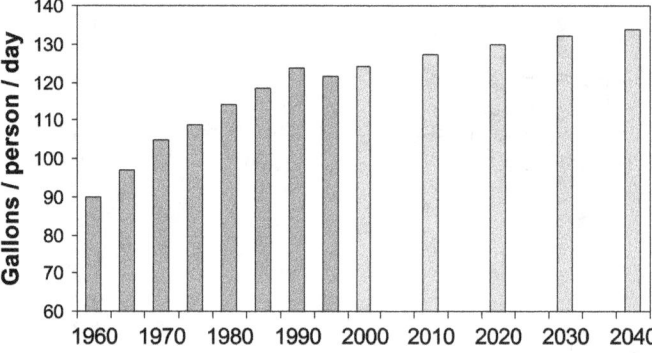

Figure 23. Alternative projections of per-capita domestic and public withdrawal.

domestic and public withdrawal, which was projected under the best-guess scenario to remain constant at the 1990-1995 average of 121 gallons per person per day. Daily per-capita withdrawal would have to decrease from 121 to 86 gallons in 2040 (29% drop) to keep total domestic and public withdrawal at the 1995 level of 32 bgd.

In the livestock industry, year 2040 daily per-capita withdrawal would have to drop from the projected 21 to 15 gallons (29% drop) to keep total livestock withdrawals at the 1995 level of 5.5 bgd (table 11). Industrial and commercial withdrawals in year 2040 would have to drop from the projected 3.89 gallons per $1000 of income to 3.63 gallons (7% drop) to keep withdrawal at the 1995 level of

Table 9. Percentage change in total withdrawals from 1995 to 2040 under best-guess scenario and if irrigated acreages of selected Western WRRs were to remain constant at their 1995 levels.

Water resource region	Best-guess scenario	Constant acreage
11. Arkansas-White-Red	–5	3
12. Texas-Gulf	6	10
13. Rio Grande	–25	–11
16. Great Basin	9	23
17. Pacific Northwest	0	16
18. California	3	6
20. Hawaii	5	21
United States	7	10

Table 10. Sensitivity of U.S. withdrawals in 2040 to assumptions about future per-capita domestic and public withdrawal.

		2040		
	1995	Low	Best guess	High
Withdrawal/person/day (gallons)	121	115	122	134
U.S. domestic & public withdrawal (bgd)	32	42	45	49
Total U.S. withdrawal (bgd)	340	361	364	368

Table 11. Required levels of water-use factors in 2040 to keep that year's total U.S. withdrawal for respective water-use categories at 1995 levels.

Water use category	Factor	Factor level		
		1995	2020	
			Best guess	Requirement
Livestock	Gallons/day/person	21	21	15
Domestic & public	Gallons/day/person	121	121	86
Industrial & commercial	Gallons/$1000	7.35	3.89	3.63
Thermoelectric	Total kWh/year/person	11,775	13,125	12,123
Thermoelectric	Gallons/kWh	23.1	15.0	13.9

36 bgd. Two options are offered for thermoelectric withdrawals. To keep thermoelectric withdrawals at their 1995 level of 132 bgd, annual per-capita electricity production would have to drop from the projected 13,125 to 12,123 kWhs (8% drop), or freshwater withdrawal per kilowatt hour produced would have to drop from the projected 15.0 to 13.9 gallons (7% drop). Some combination of these two options for thermoelectric water use could also keep withdrawals from rising.

Comparison with Past Projections

Having established the best-guess projections, let us compare them with previous projections of U.S. water withdrawal. Table 12 lists the six major projections of U.S. water withdrawals published over the past 40 years, plus those of the current study[18]. Given that actual withdrawal in 1995 was 340 bgd, this study's projection of 341 bgd for year 2000 is realistic, indicating that most previous projections seriously over-estimated U.S. withdrawal. Three of the year 2000 projections exceed 800 bgd, with a fourth above 500 bgd. The remaining two projections, by the Water Resources Council in 1978 and Guldin, are under 400 bgd.[19]

Population is a key variable of the various withdrawal projections. As figure 24 shows, studies with the highest population projections also produced the highest with-

---

[18] *The geographic areas represented by these different estimates differ slightly from one another. For example, those by the Senate Select Committee (1961) and Wollman and Bonem (1971) are for the 48 contiguous states only, and the 2 by the Water Resources Council are for the 50 states plus Puerto Rico. The current estimates are for the 50 states.*

[19] *In 1981, the USDA Forest Service published its first assessment of the nation's forest and range land situation. This assessment included projections of water withdrawals, but because they merely extended those of the Water Resources Council (1978), they are not included here.*

Table 12. Projections of U.S. water withdrawals for three future years based on medium or best-guess assumptions (bgd).

	2000	2020	2040
Senate Select Committee (1961)	888		
Water Resources Council (1968)	804	1368	
Wollman and Bonem (1971)	563	897	
National Water Commission (1973)	1000		
Water Resources Council (1978)	306		
Guldin (1989)	385	461	527
Current study	341	350	364

drawal projections. However, variations among the projections are not due to population alone (for example, the withdrawal projection of the National Water Commission nearly doubled that of Wollman and Bonem although their population projections were similar). Other major reasons for the difference include different assumptions

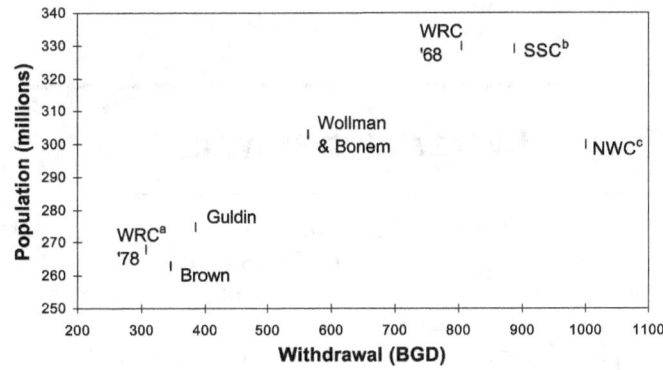

a Water Resources Council
b Senate Select Committee
c National Water Commission

Figure 24. National withdrawal and related population projections for year 2000.

about future irrigated acreage and about efficiency of industrial water use.

In comparison with the first four projection studies, the Water Resource Council's 1978 projection, now over 20 years old, was surprisingly accurate. This withdrawal projection was much lower than previous ones largely because of:

- a lower population projection,

- additional withdrawal data showing improving efficiency of industrial and thermoelectric water use, and

- knowledge of the Clean Water Act amendments passed in 1972.

The Clean Water Act, as mentioned above, was an important catalyst for additional significant improvements in water-use efficiency in the manufacturing and thermoelectric power sectors.

The Council's relatively minor under-estimation of year 2000 withdrawals resulted from a combination of under-estimating year 2000 population and over-estimating projected improvements in water-use efficiency in manufacturing and thermoelectric energy generation. As noted, efficiency improvements in the industrial and commercial and thermoelectric power sectors have been leveling off in recent years.

Also shown in table 12 is a comparison of Guldin's projections with this study's for years 2020 and 2040. Guldin's withdrawal projections are considerably higher despite his lower population projection (e.g., Guldin assumed a year 2040 U.S. population of 333 million versus this study's 370 million). Projections of the two studies differ mainly because Guldin assumed no further gains in water-use efficiency beyond those already achieved by 1985, whereas this study extrapolated trends in efficiency gains into the future.

Groundwater Withdrawals

Groundwater pumping in the U.S., as reported in the USGS water use circulars, grew by 66% from 1960 to 1995 (table 13). However, the proportion of total withdrawal coming from groundwater has remained quite stable over the past 35 years, ranging from 21% to 24%. In 1995, 22% of freshwater withdrawals came from groundwater (table 13).

Groundwater pumping increased consistently in the East and West from 1960 to 1980 (figure 25; the East consists of WRRs 1 through 9, the West consists of the remaining WRRs). The drop in 1985 shown in figure 25 for

Table 13. Groundwater withdrawal trends.

Water resource region	Withdrawal percent ground water in 1995	Percent change in withdrawal 1960–1995	
		Ground water	Surface water
1. New England	20	107	13
2. Mid Atlantic	12	57	29
3. South Atlantic-Gulf	22	163	110
4. Great Lakes	5	33	12
5. Ohio	7	49	20
6. Tennessee	3	−30	22
7. Upper Mississippi	11	98	114
8. Lower Mississippi	46	606	349
9. Souris-Red-Rainy	46	161	19
10. Missouri Basin	26	213	96
11. Arkansas-White-Red	46	148	234
12. Texas-Gulf	34	−36	129
13. Rio Grande	29	17	427
14. Upper Colorado	2	−78	473
15. Lower Colorado	38	1	−31
16. Great Basin	27	77	−10
17. Pacific Northwest	17	49	45
18. California	40	46	119
19. Alaska	27	131	−12
20. Hawaii	51	−11	−20
United States	22	66	61

the East and West is related to the overall drop in withdrawal (figure 5), and could be partly due to the noted change in data gathering procedures initiated by the USGS for the 1985 circular. Since 1985, groundwater pumping in the West has been quite stable, though it has been increasing in the East. See table A1.8 for groundwater withdrawal trends by WRR.

Over the past 35 years, groundwater withdrawal as a percent of total freshwater withdrawal has generally been falling in the West but rising in the East (figure 26). From 1960 to 1995, percent of withdrawal coming from groundwater fell from 35% to 30% in the West but rose from 9% to 15% in the East.

The regional data show that pumping has risen most quickly on a percentage basis in the following five WRRs: 3 (South Atlantic-Gulf), 8 (Lower Mississippi), 9 (Souris-Red-Rainy), 11 (Arkansas-White-Red), and 19 (Alaska). The most dramatic increase, over 600% from 1960 to 1995, occurred in WRR 8, where pumping is still climbing (table A1.8).

Although the projected future withdrawals presented in this report were not allocated to surface and groundwater sources, continued reliance on groundwater pumping would probably be essential for the projected withdrawal increases to occur. This continued pumping would in-

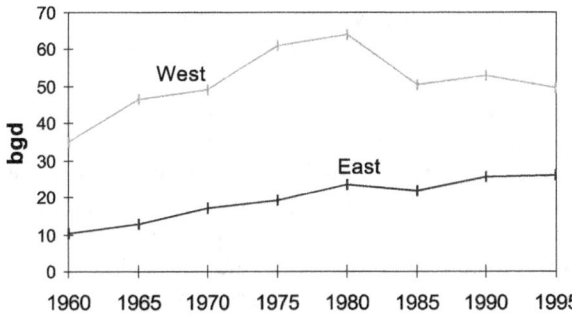

Figure 25. Groundwater withdrawal in the U.S.

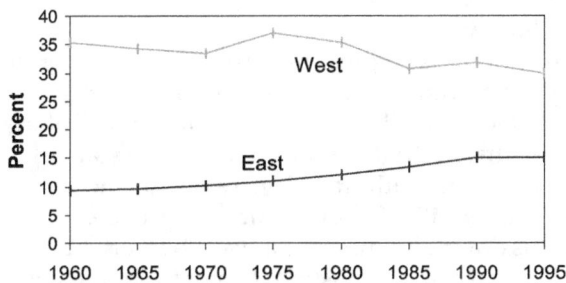

Figure 26. Portion of U.S. freshwater withdrawals from groundwater.

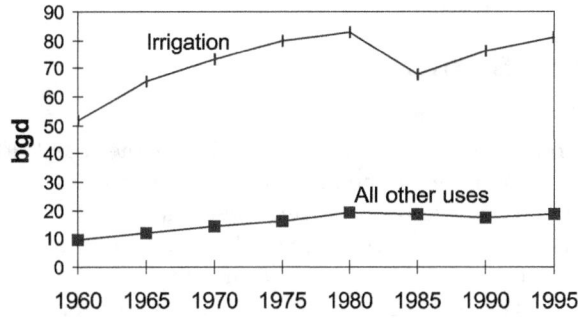

Figure 27. Consumptive use in the U.S., 1960 to 1995.

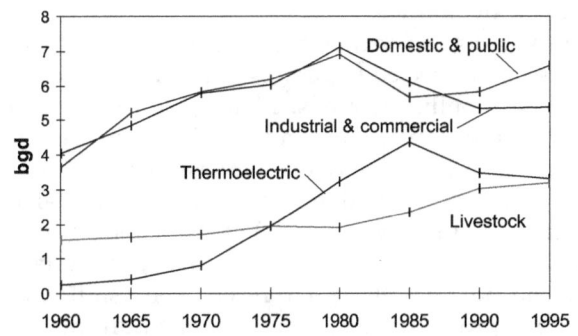

Figure 28. Consumptive use in the U.S., lessor categories.

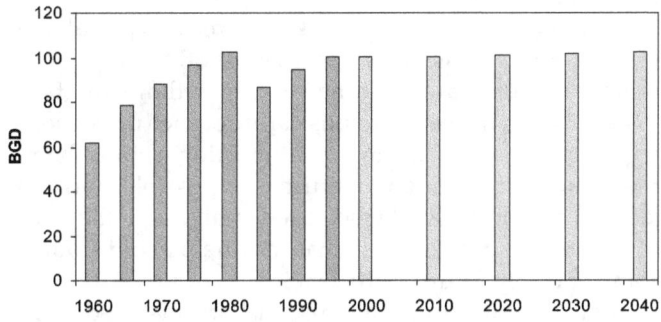

Figure 29. Projected consumptive use in the U.S.

crease pumping depths and thus, the cost of pumping, which would place additional pressure on surface water supplies.

Consumptive Use

Consumptive use is the portion of a withdrawal that does not return to the stream (it is removed from further use in the basin during the current hydrologic cycle). Most consumptive use is due to evaporation, but diversions outside the basin, or incorporated in products consumed outside the basin, are other components. Past consumptive use estimates reported here were taken from the USGS water use circulars.

The portion of withdrawal that was consumptively used in 1995 was about 60% for irrigation and for livestock, 21% for domestic and public use, 15% for industrial and commercial use, and 3% for thermoelectric power (table A1.9).

As with withdrawal, consumptive use steadily increased from 1960 to 1980, followed by a drop in 1985. Consumptive use in irrigated agriculture has been much greater than that in the other categories for the entire 35-year period (figure 27). Consumptive use in the other categories has, as a group, been quite constant since 1980 but has climbed again in agriculture.

A closer look at consumptive use in the other uses (figure 28) indicates that the constant level of consumptive use since 1980 masks substantial shifts in the individual uses. Thermoelectric consumptive use has recently been dropping, industrial and commercial consumptive use has stabilized, and consumptive use in domestic and public and livestock uses continues to climb.

Figure 29 shows projected consumptive use for the entire U.S. The projection assumes that the 1995 consumptive-use factors for each water-use category in each WRR (listed in table A1.9) apply throughout 1995 to 2040. Total consumptive use is projected to increase by 3%, from 100 bgd in 1995 to 102 bgd in 2040 (table A1.10).

Conclusions

This report projects water use in the U.S. to the year 2040 based on forecasts of population and income and on extending past trends in several key factors affecting water use. The overall approach has been to avoid complicated models, which tend to rely on arrays of obscured assumptions, and use instead a relatively straightforward extension of past trends that maintains the visual continuity of the trend, as long as those trends are sensible in light of available evidence.

In most cases, future rates of change in factors affecting water use are projected to diminish over time, in keeping with recent nonlinear trends. This is the case whether those rates are positive and thus tend to increase withdrawal, such as per-capita electricity use, or negative and thus tend to decrease withdrawal, such as industrial and commercial withdrawal per dollar or withdrawal per kilowatt per hour. In some cases, such as per-capita domestic and public withdrawal, recent trend shifts make projections problematic and a constant future rate was assumed.

Based on the best available estimates of future population, total U.S. withdrawals are projected to rise by only 7% from 1995 to 2040, despite a 41% increase in population (figure 16.1). This is in contrast to most other projections, which suggested dramatic increases in withdrawal (table 12). However, the very tenuous nature of such projections, whether they are alarming or comforting, should be remembered. Predicting the future is impossible, and projections are most useful when used to highlight the kinds of changes in public policy that may help avoid troublesome or costly future inefficiencies.

All factors affecting water use that are analyzed in this report (e.g., population, per-capita domestic and public withdrawal, per-capita electricity production, irrigated acres) and some other factors affecting water use that were not analyzed (e.g., changes in hydropower capacity) are subject to public policy. For example, population is subject to immigration policy; per-capita domestic and public withdrawal and electric energy production are subject to water or electricity prices and to regulations affecting the efficiency of water- or power-using appliances; and irrigated acreage is subject to a host of influences including subsidization of water supply, crop price supports, international trade policy, and regulations affecting the ease with which water trades may occur. Policy changes alter water use in ways not captured by extending past trends in water-use efficiencies as done herein. Indeed, the purpose of this report is to highlight areas where policy changes should be considered. Four policy implications will be offered.

First, improvements in water-use efficiency have had a positive effect on withdrawals, holding use considerably below what was projected by earlier studies that ignored changing efficiencies. Further improvements in water-use efficiency must be encouraged or at least not discouraged. Additional improvements are likely to follow the kind of influences that encouraged past improvements, such as environmental pollution regulations, price increases, reductions in government subsidies, and plumbing fixture ordinances. The cautiously optimistic projections presented herein rely critically on continued improvements in efficiencies in industry, thermoelectric power, and irrigation, and on containing growth in per-capita domestic and public withdrawal.

Second, we must strive to remove remaining barriers to voluntary water trades, allowing the market to reallocate water to its most valuable uses. This will be particularly important in the West, where water is relatively scarce. For example, projected withdrawal increases in the Upper Colorado Basin will, if they occur, force decreases in withdrawals downstream, especially in Southern California, and may eventually constrain Upper Colorado Basin withdrawals. Much of the projected increase in withdrawal in the Upper Basin would result from increased irrigation, and much of the irrigation in that region is used to grow relatively low-valued feed grains and forage crops. Opportunities for trade are ample.

Third, given the large increase in withdrawal projected for areas of the South (especially WRRs 3 and 8), and the probability that much of the projected withdrawal increases will come from increased groundwater pumping, states in these regions will probably be facing greater conflicts between groundwater and surface-water supplies wherever groundwater pumping affects surface-water flow. As conflicts become more common, water law and policy must reconsider the conjunctive nature of these two sources of water, and the saltwater intrusion problem in some coastal areas of the South.

Fourth, meeting the offstream water needs of a growing population will decrease streamflows, leading to greater environmental conflicts, especially in the West but also in some Eastern areas. Law and policy will be forced to respond to deepening concerns about instream flows.

As presented in tables 8 through 10, the best-guess projections of this report can be substantially altered by changes in the basic assumptions on which the projections are based. For example, if population is 10% above that projected, withdrawals in 2040 will increase by 1% (WRR 14) to 11% (WRR 6) above the best-guess projections. Any or all of the assumptions made in this report about factors affecting projected withdrawals may be optimistic. Further, factors ignored in this report (e.g., effects of international demand for grain on irrigation) may play a significant role in increasing demand for withdrawals. In addition, this study has ignored the problem of climatic variability (e.g., droughts). The projections apply to the average year, not to dry years, and to large-scale regions, not to specific locations that experience above average impacts. The considerable likelihood that, especially during dry years, the projected water-use increases will not be met or will be met only by causing serious side effects, suggests that it is prudent of U.S. water policy makers to encourage water conservation and improvements in the efficiency of water use when such changes can be accomplished without exorbitant cost. If the worst never happens, the unexpected beneficiaries of the policy changes may be the health and beauty of riparian ecosystems.

Finally, consider the ever-present tradeoff between population and economic growth, and ecosystem health. All else equal, a given improvement in the efficiency of water use can enable additional consumptive use or improved instream flow, but not both. In light of expected population and income growth, instream flow will only be maintained by the conscious efforts of policy makers to facilitate instream flow protection.

Acknowledgments

I am grateful to Wayne Solley and others at the USGS for their help in accessing and understanding their data, and to Michelle Haefele of Colorado State University for her assistance in data organization and analysis. Thanks also to Wayne Solley, Neil Grigg, and Linda Langner for helpful reviews of the draft.

Literature Citations

Army Corps of Engineers and Federal Emergency Management Administration. 1992. National inventory of dams database. Federal Emergency Management Administration, Washington, D.C.

Brown, T. C., B. L. Harding, and E. A. Payton. 1990. Marginal economic value of streamflow: a cast study for the Colorado River Basin. Water Resources Research 26: 2845-2859.

Bureau of the Census, U.S. Department of Commerce. 1976. Historical statistics of the United States, part 1, series J, 92-103. Washington D.C.

Bureau of the Census, U.S. Department of Commerce. 1992. USA counties on CD-Rom. Washington D.C.

Bureau of Economic Analysis, U.S. Department of Commerce. 1992. County projections to 2040 (on disk). Washington D.C.

Bureau of Economic Analysis, U.S. Department of Commerce. 1993. Regional economic information system 1969-1991 (CD-ROM). Washington D.C.

Bureau of Economic Analysis, U.S. Department of Commerce. 1995. BEA regional projections to 2045: Volume 1, states. U.S. Government Printing Office, Washington D.C.

Council on Environmental Quality. 1989. Environmental trends. U.S. Government Printing Office, Washington, D.C.

David, E. L. 1990. Manufacturing and mining water use in the United States, 1954-83. In: National water summary 1987 hydrologic events and water supply and use. Carr, J.E., E.B. Chase, R.W. Paulson and D.W. Moody, compilers. U.S. Geological Survey Water Supply Paper 2350.

Diaz, G. E., and T. C. Brown. 1997. AQUARIUS: a modeling system for river basin water allocation. General Technical Report RM-299, Rocky Mountain Forest and Range Experiment Station, Fort Collins, Colorado.

Espey, M., J. Espey, and W. D. Shaw. 1997. Price elasticity of residential demand for water: a meta-analysis. Water Resources Research 33: 1369-1374.

Gillilan, D. M., and T. C. Brown. 1997. Instream flow protection: seeking a balance in Western water use. Island Press, Washington, D.C.

Guldin, R.W. 1989. An analysis of the water situation in the United States: 1989–2040. General Technical Report RM-177, Rocky Mountain Forest and Range Experiment Station, Fort Collins, Colorado.

Forest Service, U.S. Department of Agriculture. 1981. An assessment of the forest and range land situation in the United States. Forest Resource Report No. 22, Washington, D.C.

Harding, B. L., T. B. Sangoyomi, and E. A. Payton. 1995. Impacts of a severe sustained drought on Colorado River water resources. Water Resources Bulletin 31: 815-824.

MacDonnell, D. H. Getches, and W. C. Hugenberg, Jr. 1995. The law of the Colorado River: coping with severe sustained drought. Water Resources Bulletin 31: 825-836.

MacKichan, K. A. 1951. Estimated water use in the United States, 1950. USGS Circular 115, U.S. Geological Survey, Washington D.C.

MacKichan, K. A. 1957. Estimated water use in the United States, 1955. USGS Circular 398, U.S. Geological Survey, Washington D.C.

MacKichan, K. A. and J. C. Kammerer. 1961. Estimated use of water in the United States, 1960. USGS Circular 456, U.S. Geological Survey, Washington D.C.

Moore, M. R., W. M. Crosswhite, and J. E. Hosteler. 1990. Agricultural water use in the United States, 1950-85. In: National water summary 1987 - hydrologic events and water supply and use. Carr, J. E., E. B. Chase, R. W. Paulson and D. W. Moody, compilers. U.S. Geological Survey Water Supply Paper 2350.

Murray, C. R. 1968. Estimated use of water in the United States, 1965. USGS Circular, 556, U.S. Geological Survey, Washington D.C.

Murray, C. R., and E. B. Reeves. 1972. Estimated use of water in the United States in 1970. USGS Circular 676, U.S. Geological Survey, Washington D.C.

Murray, C. R., and E. B. Reeves. 1977. Estimated use of water in the United States in 1975. USGS Circular 765, U.S. Geological Survey, Washington D.C.

National Water Commission. 1973. Water policies for the future Final report to the Congress of the United States by the National Water Commission. U.S. Government Printing Office, Washington, D.C.

Osborn, C. T., J. E. Schefter, and L. Shabman. 1986. The accuracy of water use forecasts: evaluation and implications. Water Resources Bulletin 22(1): 101-109.

Saliba, B. C. 1987. Do water markets "work"? Market transfers and trade-offs in the Southwestern states. Water Resources Research 23(7): 1113-1122.

Schefter, J. E. 1990. Domestic water use in the United States, 1960-85. In: National water summary 1987 - hydrologic events and water supply and use. Carr, J. E., E. B. Chase, R. W. Paulson and D. W. Moody, compilers. U.S. Geological Survey Water Supply Paper 2350.

Senate Select Committee on National Water Resources. 1961. Report of the Select Committee on National Water Resources pursuant to Senate Resolution 48, 86th Congress, together with suplemental and individual views: U.S. 87th Congress, 1st session, Senate Report 29.

Shabman, L. 1990. Water use forecasting - benefits and capabilities. In: National water summary 1987–hydrologic events and water supply and use. Carr, J. E., E. B. Chase, R. W. Paulson and D. W. Moody, compilers. U.S. Geological Survey Water Supply Paper 2350.

Solley, W. B., E. B. Chase, and W. B. Mann IV. 1983. Estimated use of water in the United States in 1980. USGS Circular 1001, U.S. Geological Survey, Washington D.C.

Solley, W. B., C. F. Merk, and R. R. Pierce. 1988. Estimated use of water in the United States in 1985. USGS Circular 1004, U.S. Geological Survey, Washington D.C.

Solley, W. B., R. R. Pierce, and H. A. Pearlman. 1993. Estimated use of water in the United States in 1990. USGS Circular 1081, U.S. Geological Survey, Washington D.C.

Solley, W. B., R. R. Pierce, and H. A. Pearlman. 1998. Estimated use of water in the United States in 1995. USGS Circular 1200, U.S. Geological Survey, Washington D.C.

Stockton, C. W., and G. C. Jacoby. 1976. Long-term surface-water supply and streamflow trends in the Upper Colorado River Basin based on tree-ring analyses. Lake Powell Research Project Bulletin, No. 18, National Science Foundation.

Viessman, W., Jr., and C. DeMoncada. 1980. State and national water-use trends to the year 2000; U.S. 96th Congress, 2nd session, Senate Committee on Environment and Public Works, Committee Print 96-12.

Water Resources Council. 1968. The nation's water resources. U.S. Government Printing Office, Washington, D.C.

Water Resources Council. 1978. The nation's water resources 1975–2000. U.S. Government Printing Office, Washington, D.C.

Williams, M., and B. Suh. 1986. The demand for urban water by customer class. Applied Economics 18: 1275-1289.

Wollman, N., and G. W. Bonem. 1971. The outlook for water quality, quantity, and national growth. Johns Hopkins, Baltimore, MD.

Appendix 1: Withdrawal and Consumptive Use by Decade, Past and Future

This appendix lists freshwater withdrawal and consumptive use by water resource region (WRR) for past (1960 to 1990) and future (2000 to 2040) years by decade. Past estimates are from U.S. Geological Survey (USGS) water-use circulars.

Projections were computed for this report, as described above. Tables A1.1 to A1.5 list withdrawals for each of the five water-use categories. Table A1.6 lists total withdrawal. Table A1.7 lists per-capita withdrawals. Table A1.8 lists past groundwater withdrawals. Table A1.9 lists the percent of 1995 withdrawals that the USGS estimates were consumptively used. Table A1.10 lists consumptive use.

Table A1.1 Livestock withdrawal (billion gallons per day) .

Water resource region	1960	1970	1980	1990	2000	2010	2020	2030	2040
1. New England	0.01	0.01	0.01	0.01	0.02	0.02	0.02	0.02	0.03
2. Mid Atlantic	0.06	0.08	0.11	0.10	0.14	0.14	0.15	0.16	0.17
3. South Atlantic-Gulf	0.13	0.16	0.24	0.35	0.43	0.48	0.53	0.58	0.63
4. Great Lakes	0.09	0.09	0.08	0.09	0.07	0.07	0.08	0.08	0.09
5. Ohio	0.13	0.14	0.15	0.13	0.14	0.15	0.16	0.17	0.18
6. Tennessee	0.04	0.03	0.04	0.20	0.21	0.23	0.25	0.27	0.29
7. Upper Mississippi	0.29	0.26	0.27	0.27	0.26	0.28	0.30	0.32	0.34
8. Lower Mississippi	0.04	0.06	0.04	1.07	1.05	1.11	1.19	1.27	1.35
9. Souris-Red-Rainy	0.02	0.01	0.01	0.02	0.02	0.02	0.02	0.02	0.03
10. Missouri Basin	0.32	0.44	0.39	0.41	0.45	0.48	0.52	0.56	0.60
11. Arkansas-White-Red	0.14	0.19	0.24	0.36	0.41	0.44	0.48	0.51	0.54
12. Texas-Gulf	0.07	0.11	0.20	0.16	0.22	0.24	0.26	0.28	0.30
13. Rio Grande	0.05	0.04	0.03	0.03	0.04	0.04	0.04	0.05	0.05
14. Upper Colorado	0.01	0.02	0.09	0.12	0.06	0.07	0.07	0.08	0.09
15. Lower Colorado	0.02	0.03	0.02	0.10	0.04	0.05	0.06	0.06	0.07
16. Great Basin	0.02	0.04	0.05	0.04	0.09	0.11	0.13	0.14	0.15
17. Pacific Northwest	0.06	0.05	0.06	0.62	1.60	1.79	1.98	2.15	2.31
18. California	0.08	0.09	0.09	0.41	0.48	0.54	0.59	0.64	0.69
19. Alaska	0.00	0.00	0.00	0.00	0.00	0.00	0.00	0.00	0.00
20. Hawaii	0.00	0.01	0.01	0.01	0.01	0.01	0.01	0.01	0.02
United States	1.59	1.85	2.12	4.49	5.75	6.30	6.86	7.40	7.91

USDA Forest Service Gen. Tech. Rep. RMRS–GTR–39. 1999

39

Appendix 1 Cont'd.

Table A1.2 Domestic and public withdrawal (billion gallons per day).

Water resource region	1960	1970	1980	1990	2000	2010	2020	2030	2040
1. New England	0.74	0.89	1.13	1.17	1.14	1.22	1.32	1.42	1.51
2. Mid Atlantic	2.73	3.94	4.33	4.77	5.12	5.39	5.74	6.08	6.43
3. South Atlantic-Gulf	1.61	2.16	3.32	3.91	4.86	5.44	6.03	6.59	7.10
4. Great Lakes	2.29	2.98	2.07	3.13	3.47	3.61	3.82	4.04	4.26
5. Ohio	1.10	1.67	1.71	1.86	2.00	2.11	2.25	2.40	2.54
6. Tennessee	0.29	0.23	0.37	0.36	0.42	0.46	0.50	0.53	0.57
7. Upper Mississippi	0.81	1.20	1.40	1.19	1.21	1.28	1.37	1.46	1.55
8. Lower Mississippi	0.32	0.57	0.81	0.91	0.94	1.00	1.07	1.14	1.20
9. Souris-Red-Rainy	0.04	0.06	0.07	0.08	0.07	0.07	0.07	0.08	0.08
10. Missouri Basin	0.68	0.84	1.33	1.37	1.38	1.49	1.61	1.73	1.84
11. Arkansas-White-Red	0.53	0.61	0.92	1.00	1.11	1.19	1.28	1.38	1.46
12. Texas-Gulf	0.62	0.86	1.72	2.34	2.78	3.05	3.32	3.57	3.80
13. Rio Grande	0.24	0.22	0.33	0.47	0.44	0.49	0.53	0.58	0.62
14. Upper Colorado	0.05	0.06	0.17	0.11	0.13	0.15	0.16	0.18	0.19
15. Lower Colorado	0.23	0.35	0.53	0.83	0.99	1.15	1.30	1.43	1.56
16. Great Basin	0.27	0.31	0.69	0.52	0.51	0.60	0.68	0.75	0.82
17. Pacific Northwest	0.88	0.96	1.03	1.42	1.59	1.78	1.96	2.13	2.29
18. California	2.41	2.93	3.44	4.78	4.74	5.29	5.81	6.29	6.76
19. Alaska	0.02	0.06	0.05	0.01	0.06	0.07	0.07	0.08	0.08
20. Hawaii	0.06	0.11	0.14	0.18	0.18	0.20	0.22	0.24	0.25
United States	15.91	21.01	25.56	30.38	33.12	36.02	39.12	42.10	44.93

Table A1.3 Industrial and commercial withdrawal (billion gallons per day).

Water resource region	1960	1970	1980	1990	2000	2010	2020	2030	2040
1. New England	1.55	1.81	1.99	1.02	0.75	0.73	0.72	0.74	0.76
2. Mid Atlantic	4.77	8.30	4.90	3.86	3.12	3.00	2.92	2.97	3.05
3. South Atlantic-Gulf	3.62	4.48	7.10	4.99	4.87	5.03	5.15	5.41	5.68
4. Great Lakes	9.13	10.30	7.80	6.04	5.81	5.53	5.36	5.44	5.57
5. Ohio	7.82	6.51	5.79	4.49	4.32	4.19	4.13	4.22	4.36
6. Tennessee	1.58	1.52	2.10	1.55	1.32	1.32	1.32	1.37	1.42
7. Upper Mississippi	2.08	2.26	4.16	2.45	2.30	2.25	2.21	2.26	2.33
8. Lower Mississippi	1.52	4.15	4.51	2.97	3.12	3.08	3.05	3.14	3.25
9. Souris-Red-Rainy	0.09	0.08	0.02	0.07	0.04	0.04	0.04	0.04	0.04
10. Missouri Basin	0.60	0.82	1.00	0.87	0.87	0.87	0.87	0.90	0.93
11. Arkansas-White-Red	1.15	0.84	1.63	1.11	1.21	1.21	1.21	1.25	1.29
12. Texas-Gulf	1.41	1.77	1.92	1.18	1.59	1.62	1.63	1.68	1.75
13. Rio Grande	0.25	0.32	0.04	0.18	0.18	0.18	0.18	0.19	0.20
14. Upper Colorado	0.03	0.07	0.61	0.09	0.06	0.07	0.07	0.07	0.08
15. Lower Colorado	0.14	0.28	0.48	0.64	0.54	0.58	0.60	0.64	0.68
16. Great Basin	0.38	0.27	0.66	0.32	0.35	0.39	0.41	0.43	0.46
17. Pacific Northwest	2.49	2.38	4.20	2.14	2.82	2.91	2.96	3.10	3.25
18. California	0.84	1.08	1.24	1.70	2.27	2.32	2.34	2.44	2.55
19. Alaska	0.09	0.12	0.14	0.21	0.12	0.13	0.13	0.13	0.14
20. Hawaii	0.17	0.30	0.11	0.15	0.11	0.11	0.12	0.12	0.13
United States	39.71	47.65	50.39	36.01	35.78	35.54	35.42	36.53	37.90

40

USDA Forest Service Gen. Tech. Rep. RMRS–GTR–39. 1999

Appendix 1 Cont'd.

Table A1.4 Thermoelectric power withdrawal (billion gallons per day).

Water resource region	1960	1970	1980	1990	2000	2010	2020	2030	2040
1. New England	0.63	1.94	2.22	2.40	1.68	1.71	1.75	1.79	1.83
2. Mid Atlantic	8.10	15.56	14.33	12.20	12.57	12.43	12.50	12.63	12.78
3. South Atlantic-Gulf	8.42	15.23	19.30	19.74	18.11	19.06	20.03	20.85	21.55
4. Great Lakes	17.50	25.52	27.17	22.80	22.57	22.20	22.33	22.62	22.97
5. Ohio	16.02	27.01	30.50	23.87	22.45	22.17	22.36	22.67	23.00
6. Tennessee	5.60	6.10	9.22	7.07	7.15	7.43	7.75	8.04	8.31
7. Upper Mississippi	8.21	12.51	16.09	16.52	19.04	19.01	19.23	19.50	19.80
8. Lower Mississippi	0.95	4.13	7.74	5.65	6.72	6.71	6.80	6.89	7.00
9. Souris-Red-Rainy	0.00	0.14	0.05	0.03	0.04	0.04	0.04	0.04	0.04
10. Missouri Basin	2.21	3.32	8.19	10.03	8.95	9.17	9.44	9.68	9.90
11. Arkansas-White-Red	3.17	1.96	2.07	4.53	4.20	4.26	4.35	4.44	4.53
12. Texas-Gulf	2.18	4.75	0.98	4.71	7.80	8.02	8.24	8.42	8.60
13. Rio Grande	0.30	0.02	0.02	0.02	0.02	0.02	0.02	0.02	0.02
14. Upper Colorado	0.12	0.10	0.14	0.18	0.15	0.16	0.17	0.18	0.18
15. Lower Colorado	0.02	0.05	0.09	0.11	0.07	0.07	0.08	0.09	0.09
16. Great Basin	0.08	0.13	0.13	0.03	0.03	0.03	0.03	0.03	0.03
17. Pacific Northwest	0.01	0.03	0.03	0.36	0.68	1.16	1.54	1.84	2.09
18. California	0.43	1.50	1.99	0.25	0.22	0.25	0.27	0.29	0.31
19. Alaska	0.09	0.07	0.03	0.03	0.03	0.04	0.04	0.04	0.04
20. Hawaii	0.03	0.13	0.14	0.10	0.07	0.07	0.08	0.08	0.08
United States	74.0	120.2	140.4	130.6	132.6	134.0	137.1	140.1	143.2

Table A1.5 Irrigation withdrawal (billion gallons per day).

Water resource region	1960	1970	1980	1990	2000	2010	2020	2030	2040
1. New England	0.01	0.08	0.05	0.12	0.13	0.13	0.13	0.13	0.13
2. Mid Atlantic	0.08	0.13	0.25	0.20	0.25	0.26	0.27	0.28	0.28
3. South Atlantic-Gulf	0.80	2.40	3.80	4.70	4.71	5.07	5.35	5.57	5.74
4. Great Lakes	0.05	0.09	0.33	0.29	0.32	0.35	0.36	0.38	0.39
5. Ohio	0.01	0.04	0.15	0.07	0.12	0.15	0.17	0.18	0.19
6. Tennessee	0.01	0.01	0.01	0.03	0.04	0.05	0.06	0.07	0.07
7. Upper Mississippi	0.04	0.10	0.38	0.39	0.48	0.53	0.56	0.59	0.62
8. Lower Mississippi	0.85	3.20	7.70	7.38	8.81	10.12	11.16	11.95	12.54
9. Souris-Red-Rainy	0.01	0.01	0.06	0.10	0.11	0.12	0.13	0.13	0.14
10. Missouri Basin	13.27	18.59	29.00	24.80	24.32	24.16	24.03	23.92	23.84
11. Arkansas-White-Red	4.62	8.21	10.82	8.40	8.67	8.27	7.96	7.71	7.51
12. Texas-Gulf	7.16	6.11	5.56	5.13	4.94	4.71	4.53	4.39	4.28
13. Rio Grande	5.28	5.52	4.30	5.29	4.98	4.67	4.43	4.25	4.10
14. Upper Colorado	5.96	7.85	7.48	6.59	7.92	8.40	8.74	8.96	9.10
15. Lower Colorado	5.14	6.51	7.61	6.24	6.26	6.22	6.19	6.16	6.14
16. Great Basin	5.21	5.91	5.90	6.35	5.95	5.67	5.45	5.28	5.15
17. Pacific Northwest	18.90	27.01	29.12	31.76	26.36	24.50	23.32	22.54	22.03
18. California	16.43	34.12	38.15	28.43	28.96	28.42	27.99	27.64	27.36
19. Alaska	0.00	0.00	0.00	0.00	0.00	0.00	0.00	0.00	0.00
20. Hawaii	0.92	1.29	0.91	0.76	0.73	0.68	0.64	0.61	0.59
United States	84.8	127.2	151.6	137.0	134.1	132.5	131.5	130.7	130.2

USDA Forest Service Gen. Tech. Rep. RMRS–GTR–39. 1999

41

Appendix 1 Cont'd.

Table A1.6 Total withdrawal (billion gallons per day).

Water resource region	1960	1970	1980	1990	2000	2010	2020	2030	2040
1. New England	2.94	4.72	5.40	4.73	3.72	3.81	3.94	4.10	4.26
2. Mid Atlantic	15.75	28.01	23.92	21.12	21.19	21.22	21.59	22.12	22.71
3. South Atlantic-Gulf	14.58	24.44	33.76	33.68	32.98	35.07	37.10	39.01	40.70
4. Great Lakes	29.06	38.97	37.46	32.35	32.24	31.75	31.95	32.55	33.29
5. Ohio	25.08	35.37	38.30	30.42	29.03	28.78	29.08	29.65	30.28
6. Tennessee	7.52	7.89	11.73	9.21	9.14	9.50	9.89	10.28	10.67
7. Upper Mississippi	11.43	16.33	22.30	20.82	23.29	23.35	23.68	24.14	24.64
8. Lower Mississippi	3.68	12.10	20.79	17.97	20.64	22.02	23.27	24.39	25.33
9. Souris-Red-Rainy	0.17	0.31	0.22	0.29	0.27	0.28	0.30	0.31	0.33
10. Missouri Basin	17.08	24.01	39.91	37.49	35.96	36.17	36.47	36.78	37.10
11. Arkansas-White-Red	9.61	11.81	15.67	15.39	15.60	15.38	15.28	15.28	15.34
12. Texas-Gulf	11.44	13.61	10.38	13.51	17.33	17.64	17.98	18.34	18.72
13. Rio Grande	6.12	6.12	4.72	6.00	5.65	5.40	5.21	5.08	4.99
14. Upper Colorado	6.17	8.10	8.49	7.08	8.32	8.85	9.21	9.47	9.64
15. Lower Colorado	5.55	7.21	8.72	7.93	7.91	8.08	8.22	8.38	8.53
16. Great Basin	5.95	6.66	7.42	7.25	6.93	6.80	6.69	6.64	6.62
17. Pacific Northwest	22.34	30.42	34.43	36.29	33.05	32.14	31.77	31.76	31.97
18. California	20.19	39.72	44.91	35.55	36.67	36.81	37.00	37.30	37.66
19. Alaska	0.20	0.25	0.23	0.24	0.22	0.23	0.24	0.25	0.26
20. Hawaii	1.18	1.83	1.31	1.20	1.10	1.07	1.06	1.06	1.06
United States	216.0	317.9	370.1	338.5	341.3	344.3	349.9	356.9	364.1

Table A1.7 U.S. per-capita withdrawals (gallons per day).

	Livestock	Domestic & public	Industrial & commercial	Thermoelectric	Irrigation	Total
1960	9	90	224	417	478	1217
1970	9	104	237	597	631	1579
1980	9	114	225	625	675	1649
1990	18	123	147	529	553	1370
2000	21	122	134	486	491	1254
2010	21	122	123	453	448	1167
2020	21	122	113	427	410	1094
2030	21	122	109	407	379	1038
2040	21	122	106	389	354	992

Appendix 1 Cont'd.

Table A1.8 Groundwater withdrawal from 1960 to 1995 (billion gallons per day).

Water resource region	1960	1970	1980	1990	1995
1. New England	0.35	0.63	0.65	0.69	0.73
2. Mid Atlantic	1.68	2.60	2.40	2.64	2.63
3. South Atlantic-Gulf	2.70	4.70	6.60	7.11	7.11
4. Great Lakes	1.14	1.40	1.60	1.21	1.51
5. Ohio	1.30	1.70	2.50	2.65	1.94
6. Tennessee	0.37	0.17	0.26	0.31	0.26
7. Upper Mississippi	1.30	2.20	2.60	2.62	2.57
8. Lower Mississippi	1.30	3.60	6.70	8.34	9.18
9. Souris-Red-Rainy	0.04	0.07	0.11	0.13	0.12
10. Missouri Basin	2.98	5.90	12.00	8.49	9.32
11. Arkansas-White-Red	3.02	6.60	9.40	7.42	7.49
12. Texas-Gulf	9.35	6.20	5.10	5.48	5.96
13. Rio Grande	1.65	2.40	1.90	2.14	1.93
14. Upper Colorado	0.53	0.11	0.14	0.13	0.12
15. Lower Colorado	2.98	4.50	4.50	3.08	3.00
16. Great Basin	0.91	1.10	1.60	1.97	1.61
17. Pacific Northwest	3.70	4.30	8.20	9.78	5.50
18. California	10.00	18.00	21.00	14.40	14.57
19. Alaska	0.03	0.04	0.05	0.06	0.06
20. Hawaii	0.58	0.91	0.80	0.59	0.52
United States	45.90	67.13	88.11	79.24	76.11

Table A1.9 1995 consumptive use factors (percent).

	Livestock	Domestic & public	Industrial & commercial	Thermoelectric	Irrigation	Total
1. New England	82	13	10	1	97	10
2. Mid Atlantic	70	7	9	1	70	5
3. South Atlantic-Gulf	93	19	14	2	68	17
4. Great Lakes	78	7	9	2	94	5
5. Ohio	82	10	13	4	93	6
6. Tennessee	21	13	10	0	99	3
7. Upper Mississippi	86	28	12	2	93	7
8. Lower Mississippi	77	58	9	4	72	39
9. Souris-Red-Rainy	100	26	17	0	89	48
10. Missouri Basin	92	32	24	2	53	39
11. Arkansas-White-Red	98	35	15	4	76	51
12. Texas-Gulf	100	36	38	3	95	42
13. Rio Grande	92	41	57	79	44	44
14. Upper Colorado	23	29	35	88	33	34
15. Lower Colorado	99	43	60	90	57	57
16. Great Basin	16	34	46	97	56	54
17. Pacific Northwest	4	13	8	5	39	33
18. California	72	24	26	5	79	69
19. Alaska	98	8	13	10	53	12
20. Hawaii	47	46	39	1	63	54
United States	58	21	15	3	60	29

Appendix 1 Cont'd.

Table A1.10 Total consumptive use (billion gallons per day), assuming 1995 consumptive use factors for future years.

Water resource region	1960	1970	1980	1990	2000	2010	2020	2030	2040
1. New England	0.28	0.41	0.36	0.39	0.37	0.39	0.40	0.41	0.43
2. Mid Atlantic	1.15	1.43	1.69	1.26	1.10	1.12	1.15	1.20	1.24
3. South Atlantic-Gulf	1.97	3.26	5.13	5.14	5.58	6.02	6.42	6.77	7.08
4. Great Lakes	0.92	1.21	1.25	1.64	1.56	1.56	1.58	1.63	1.68
5. Ohio	0.81	0.93	1.67	2.11	1.83	1.85	1.89	1.95	2.01
6. Tennessee	0.41	0.24	0.37	0.32	0.29	0.31	0.33	0.34	0.36
7. Upper Mississippi	0.58	0.76	1.47	1.96	1.68	1.74	1.82	1.90	1.98
8. Lower Mississippi	1.26	3.57	6.45	6.97	8.25	9.28	10.14	10.82	11.36
9. Souris-Red-Rainy	0.06	0.07	0.13	0.15	0.14	0.15	0.16	0.17	0.18
10. Missouri Basin	7.59	12.86	16.35	12.08	14.07	14.06	14.07	14.10	14.14
11. Arkansas-White-Red	4.01	6.80	9.39	7.87	7.76	7.51	7.34	7.23	7.15
12. Texas-Gulf	5.56	6.15	6.47	5.92	6.80	6.72	6.68	6.68	6.71
13. Rio Grande	3.65	3.31	2.31	3.46	2.51	2.40	2.32	2.27	2.23
14. Upper Colorado	3.55	4.08	2.27	2.48	2.82	2.99	3.12	3.20	3.26
15. Lower Colorado	3.59	5.07	4.93	5.00	4.41	4.49	4.56	4.63	4.71
16. Great Basin	3.42	3.14	3.95	3.43	3.73	3.63	3.55	3.49	3.46
17. Pacific Northwest	8.32	10.66	11.89	12.15	10.85	10.18	9.77	9.52	9.37
18. California	13.65	22.72	25.06	20.79	25.09	24.85	24.67	24.57	24.52
19. Alaska	0.00	0.02	0.03	0.03	0.03	0.03	0.03	0.03	0.03
20. Hawaii	0.42	0.86	0.68	0.63	0.59	0.57	0.55	0.55	0.54
United States	61.2	87.6	101.8	93.8	99.5	99.9	100.6	101.5	102.4

44

USDA Forest Service Gen. Tech. Rep. RMRS–GTR–39. 1999

Appendix 2: Projected Water Use in Forest and Rangeland Renewable Resources Planning Act Assessment Regions

Resource assessments performed pursuant to the Forest and Rangeland Renewable Resources Planning Act of 1974 (RPA) typically summarize findings according to RPA assessment regions. This breakdown separates the U.S. along state lines into four regions. Figure A2.1 shows the assessment region and state boundaries, as well as the water resource region boundaries.

Summarization of the U.S. Geological Survey (USGS) data by RPA assessment region was feasible because the USGS has documented water use by state from the inception of its effort to estimate water use. Aside from the change to states as the basic geographic unit of analysis, all other methods are identical to those used to summarize and project water use for watersheds, described earlier.

The four parts of figure A2.2 depict the projected freshwater withdrawals for the four RPA assessment regions assuming the best-guess projections of water-use factors. These results are summarized in table A2.1 in terms of the change from 1995 to 2040. All four regions show substantial increases in livestock and domestic and public withdrawals, reflecting the expected population increases. Industrial and commercial withdrawals are projected to decrease slightly in the North and increase moderately in the other three regions. Percentage increases in industrial and commercial withdrawals are considerably below corresponding percentage increases in population because of the improving efficiency of industrial and commercial water use.

Thermoelectric withdrawals in the North and South are largely related to population increases; as with industrial and commercial withdrawals, they are lower in percentage terms than are the population increases because of improving efficiency of thermoelectric water use (table A2.1). In the Rocky Mountains and especially in the Pacific region, thermoelectric withdrawals in percentage terms are also affected by the relatively large proportion of electricity production currently occurring at hydroelectric plants. With the assumption of no future increases in production at hydroelectric plants, all of the expected increases in electricity production must occur at thermoelectric plants, intensifying the effect of population increases on thermoelectric withdrawals relative to eastern portions of the country.

Projected irrigation withdrawals differ considerably among regions, largely reflecting projected changes in irrigated acreage. A large drop in withdrawal is projected for the Pacific region, whereas withdrawals are expected to increase in the other three regions (table A2.1).

Total withdrawals are projected to increase in all four regions (table A2.1). The increases vary from near zero in the Pacific region to 18% in the Southern region. All of the percentage increases in withdrawals are substantially lower than the corresponding percentage increases in population. Most of the increases in the North and South are attributable to domestic and public and thermoelectric uses. Increases in the Rocky Mountains are largely attributable to irrigation in addition to domestic and public and thermoelectric uses. In the Pacific region, the large decrease in irrigation withdrawal is equaled by increases in the other four withdrawal categories.

Table A2.1 Change in population and withdrawal from 1995 to 2040 for the Forest and Rangeland Renewable Resources Planning Act assessment regions (percent change in parenthesis).

	North		South		Rocky Mountains		Pacific	
Population (millions)	34	(29%)	39	(47%)	11	(53%)	22	(53%)
Withdrawal (bgd)[a]								
Livestock	0.2	(29%)	1.0	(47%)	1.1	(53%)	0.3	(53%)
Domestic & public	3.5	(29%)	4.8	(47%)	1.7	(53%)	3.1	(53%)
Industrial & commercial	−0.9	(−5%)	1.5	(11%)	0.4	(16%)	0.6	(12%)
Thermoelectric	1.3	(2%)	8.3	(16%)	1.4	(27%)	0.7	(104%)
Irrigation	0.5	(26%)	2.3	(10%)	2.6	(4%)	−4.7	(−11%)
Total	4.6	(4%)	17.9	(18%)	7.2	(9%)	0	(0%)

[a] bgd - billion gallons per day

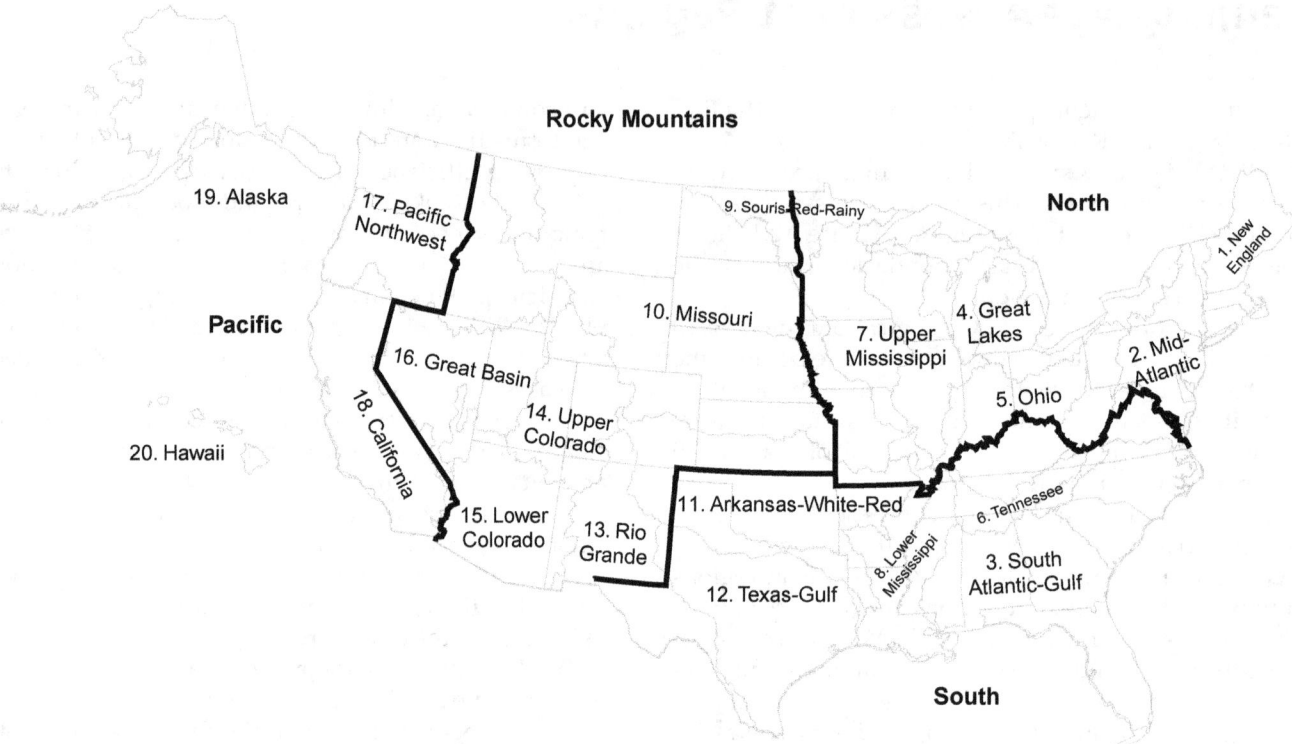

Figure A2.1. Four Forest and Rangeland Renewable Resources Planning Act assessment regions, with state and water resource region boundaries.

North

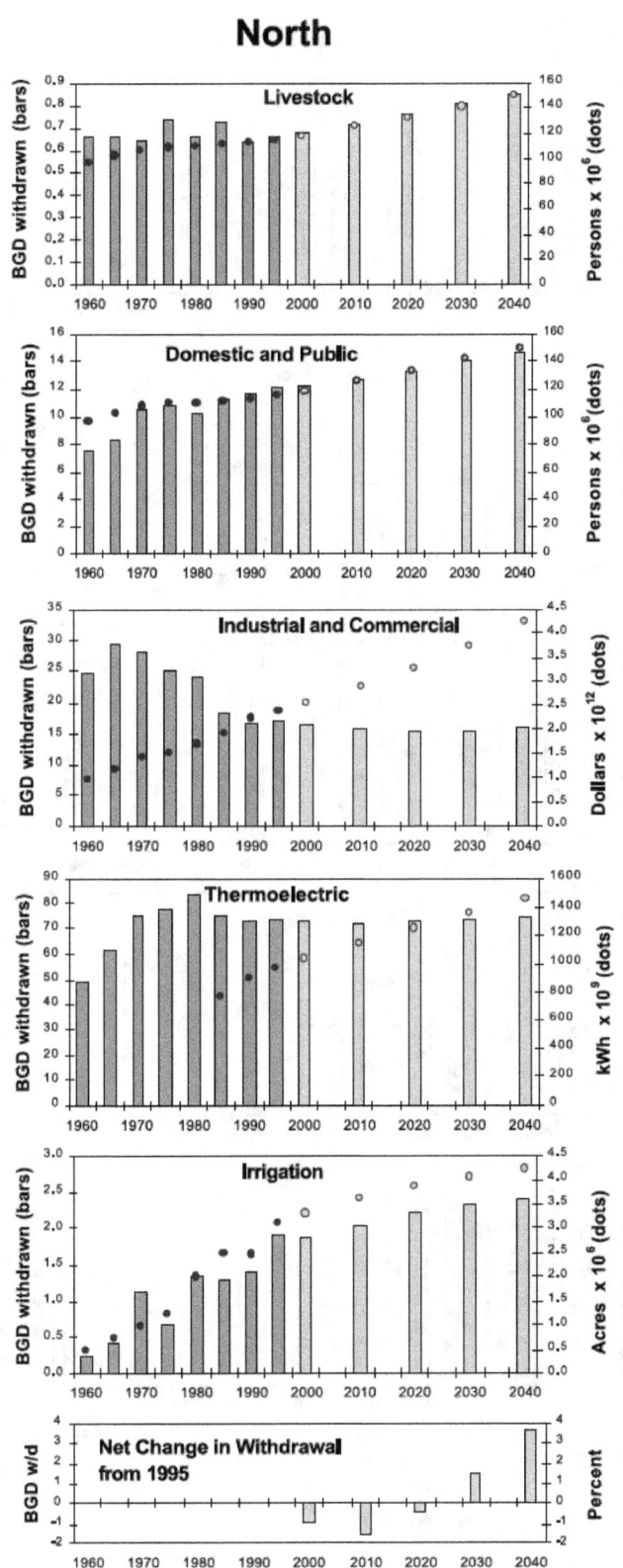

Figure A2.2.1. Withdrawal projections for RPA North assessment region.

South

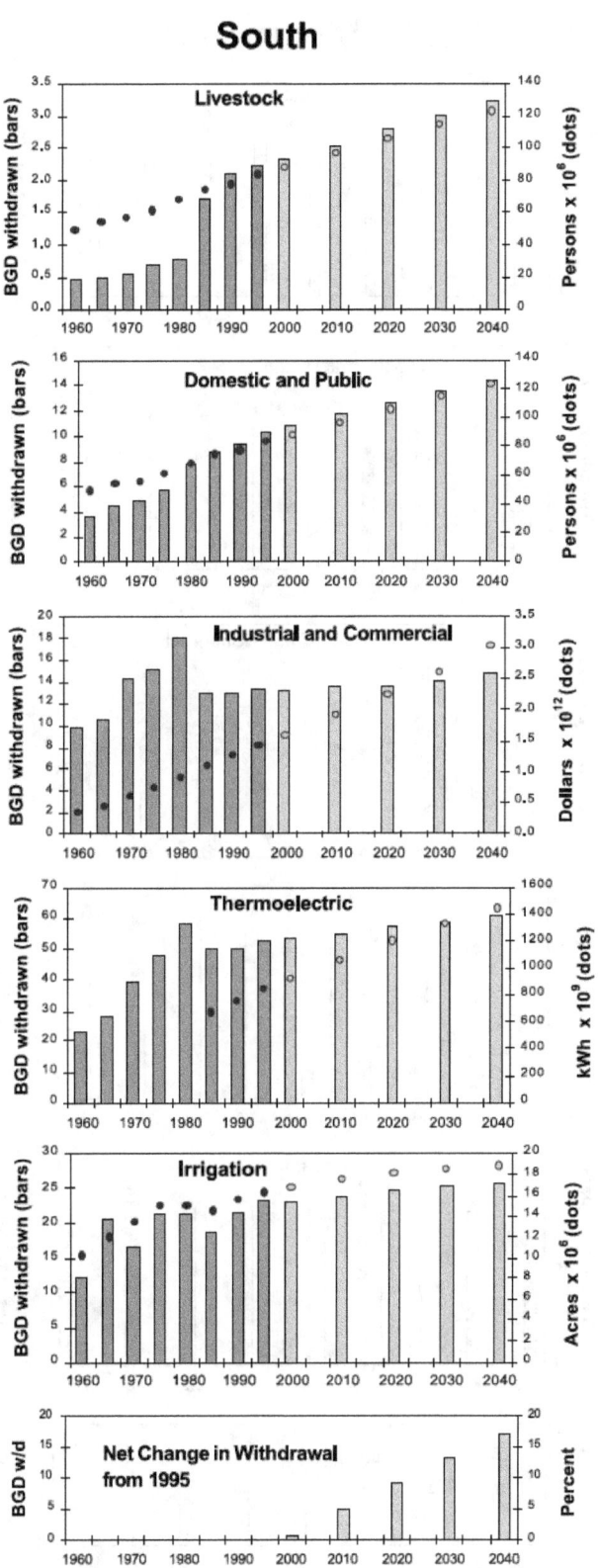

Figure A2.2.2. Withdrawal projections for RPA South assessment region.

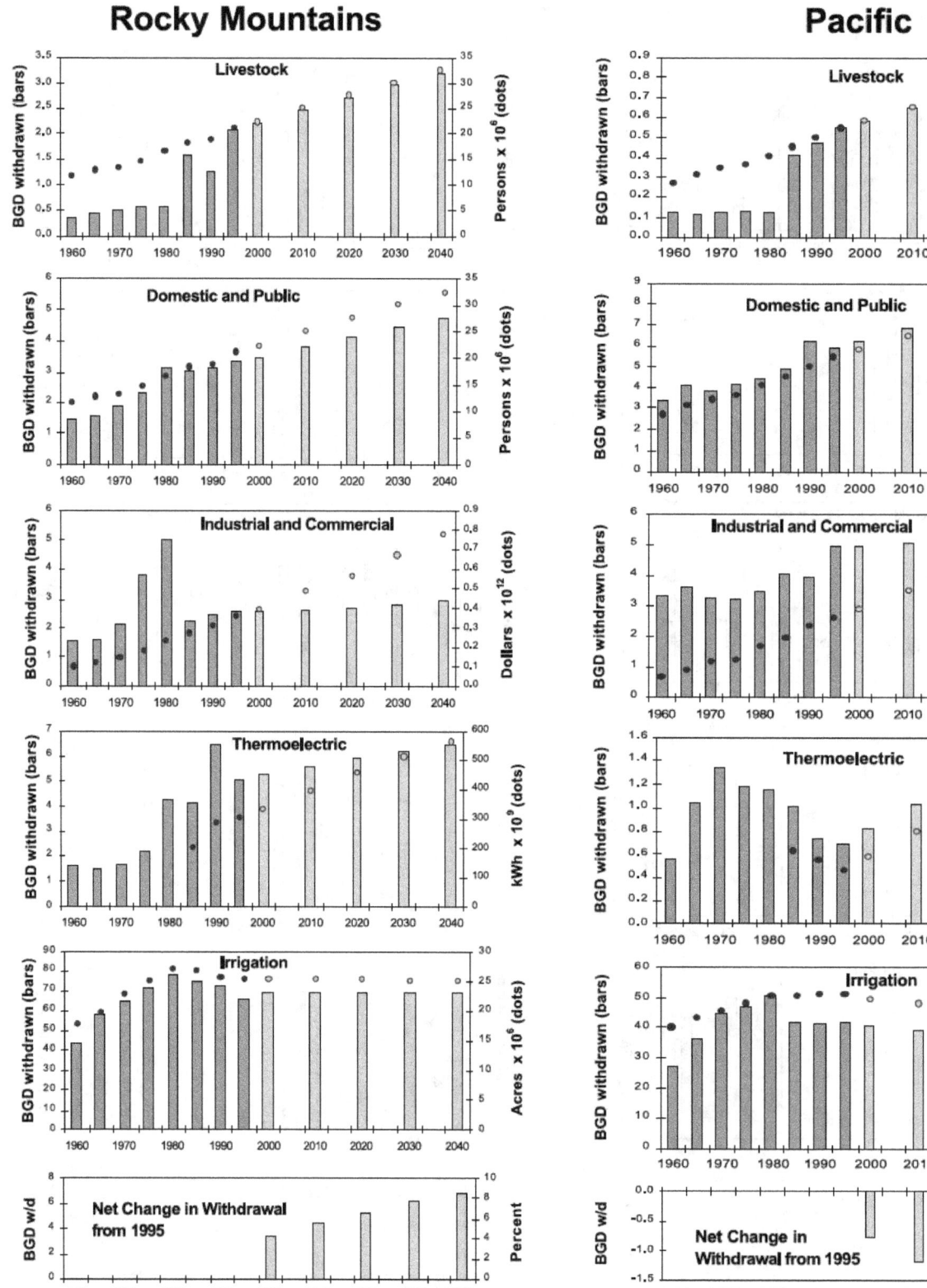

Figure A2.2.3. Withdrawal projections for RPA Rocky Mountains assessment region.

Figure A2.2.4. Withdrawal projections for RPA Pacific assessment region.

RMRS
ROCKY MOUNTAIN RESEARCH STATION

The Rocky Mountain Research Station develops scientific information and technology to improve management, protection, and use of forests and rangelands. Research is designed to meet the needs of National Forest managers, federal and state agencies, public and private organizations, academic institutions, industry, and individuals.

Studies accelerate solutions to problems involving ecosystems, range, forests, water, recreation, fire, resource inventory, land reclamation, community sustainability, forest engineering technology, multiple use economics, wildlife and fish habitat, and forest insects and diseases. Studies are conducted cooperatively, and applications can be found worldwide.

Research Locations

Flagstaff, Arizona
Fort Collins, Colorado*
Boise, Idaho
Moscow, Idaho
Bozeman, Montana
Missoula, Montana
Lincoln, Nebraska

Reno, Nevada
Albuquerque, New Mexico
Rapid City, South Dakota
Logan, Utah
Ogden, Utah
Provo, Utah
Laramie, Wyoming

* Station Headquarters, 240 West Prospect Road, Fort Collins, CO 80526

Printed on ♻ recycled paper